Kotlin Programming By Example

Build real-world Android and web applications the Kotlin way

Iyanu Adelekan

BIRMINGHAM - MUMBAI

Kotlin Programming By Example

Commissioning Editor: Aaron Lazar
Acquisition Editor: Denim Pinto
Content Development Editor: Vikas Tiwari
Technical Editor: Subhalaxmi Nadar
Copy Editor: Safis Editing
Project Coordinator: Ulhas Kambali
Proofreader: Safis Editing
Indexer: Mariammal Chettiyar
Graphics: Tania Dutta
Production Coordinator: Aparna Bhagat

First published: March 2018

Production reference: 1270318

Published by Packt Publishing Ltd.
Livery Place
35 Livery Street
Birmingham
B3 2PB, UK.

ISBN 978-1-78847-454-2

www.packtpub.com

To my mother and father for their unwavering belief in me and for the unparalleled love they have shown toward me; words cannot describe how grateful I am to you

`mapt.io`

Mapt is an online digital library that gives you full access to over 5,000 books and videos, as well as industry leading tools to help you plan your personal development and advance your career. For more information, please visit our website.

Why subscribe?

- Spend less time learning and more time coding with practical eBooks and Videos from over 4,000 industry professionals

- Improve your learning with Skill Plans built especially for you

- Get a free eBook or video every month

- Mapt is fully searchable

- Copy and paste, print, and bookmark content

PacktPub.com

Did you know that Packt offers eBook versions of every book published, with PDF and ePub files available? You can upgrade to the eBook version at `www.PacktPub.com` and as a print book customer, you are entitled to a discount on the eBook copy. Get in touch with us at `service@packtpub.com` for more details.

At `www.PacktPub.com`, you can also read a collection of free technical articles, sign up for a range of free newsletters, and receive exclusive discounts and offers on Packt books and eBooks.

Contributors

About the author

Iyanu Adelekan is a software engineer who enjoys solving problems in the web and Android application domains. He loves working on open source projects and is the author and lead maintainer of Kanary, a Kotlin web framework for building RESTful application programming interfaces. In addition to building software, he is passionate about knowledge sharing and algorithms. He enjoys reading sci-fi and playing chess in his spare time.

I am thankful to God for giving me the strength required to embark on and successfully complete this book.

Gratitude also go out to my siblings, parents, and friends for their love and support over the course of writing this title. May God bless you all.

About the reviewer

Egor Andreevici has been building Android apps since 2010 and has recently moved to Canada to join Square as an Android Developer. He is passionate about technology, clean code, test-driven development, and software architecture. He discovered Kotlin a couple of years ago and has fallen in love with conciseness and expressiveness of the language ever since. In his free time, he likes to contribute to open source, read, and travel.

Packt is searching for authors like you

If you're interested in becoming an author for Packt, please visit authors.packtpub.com and apply today. We have worked with thousands of developers and tech professionals, just like you, to help them share their insight with the global tech community. You can make a general application, apply for a specific hot topic that we are recruiting an author for, or submit your own idea.

Table of Contents

Preface

Since its announcement as an officially supported language for Android, Kotlin has experienced a huge rise in popularity. This proliferation in popularity is in many ways warranted as Kotlin is a modern and extremely well-designed language, which has many domains of application—web, mobile, and native to name a few. As a direct consequence of its rise in popularity, Kotlin has experienced steady growth over the years.

Who this book is for

This book will be useful to readers of all ages, experience levels, and demographics. That being said, it was written primarily with beginners and experienced programmers who want to learn Kotlin in mind.

Over the course of writing this book, I paid special attention to the fact that beginners will need to be eased into topics and concepts. As such, the chapters of this book were written in ascending order of difficulty. If you are a beginner, this will enable you to learn as quickly as possible with as few hiccups as possible.

Fairly experienced readers are expected to move more quickly through this book than beginners—all things being equal. If you have prior experience of programming and application development, you may choose to first run through the code examples of this book to get a good feel for the topics covered and what to expect. Java developers in particular may be able to get away with diving right into the more advanced chapters of the book.

Regardless of which category of user you fall into, rest assured that there is something that has been written specially for you!

What this book covers

`Chapter 1`, *The Fundamentals,* explains how to use Kotlin to write simple programs, how to set up a new Android project, and the fundamental knowledge required to develop Android applications that communicate with web servers.

`Chapter 2`, *Building an Android Application - Tetris,* gets you off to a quick start with Android development by walking you through the creation of the classic game, *Tetris.*

Chapter 3, *Implementing Tetris Logic and Functionality*, explains how to create views, implement application logic with models, and present data to views. In addition, you will learn about UI event handling.

Chapter 4, *Designing and Implementing the Messenger Backend with Spring Boot 2.0*, explains how to successfully design and implement a backend that will provide web resources to client applications.

Chapter 5, *Building the Messenger Android App - Part I*, covers how to build powerful Android applications with the use of the Model-View-Presenter pattern by creating a Messenger application that communicates with the Messenger backend.

Chapter 6, *Building the Messenger Android App - Part II*, continues where the previous chapter left off. This chapter completes the development of the Messenger application.

Chapter 7, *Storing Information in a Database*, explains the various methods of data storage that the Android application framework puts at your disposal. In addition, you will learn how to use these methods to store and retrieve useful application information.

Chapter 8, *Securing and Deploying the Messenger App*, gives you a step-by-step breakdown of the Android application deployment process. In addition, the chapter explains important security considerations for Android applications.

Chapter 9, *Creating the Place Reviewer Backend with Spring*, gets deeper into the exploration of the Spring Framework by design and implements a backend for a Place Reviewer web application with Spring MVC.

Chapter 10, *Implementing the Place Reviewer Frontend*, explains how to create a web app that is location-aware and utilizes the power of the Google Places API. You will also learn how to write tests for your web applications.

To get the most out of this book

If you are a beginner, begin this book with an open mind and a drive to learn. You may find the task of learning a new language daunting at first, but with perseverance, you will be a professional in no time. Read through each chapter in the order they are presented in. This will ensure you grasp all of the content within the book. Take special care to read through each code snippet so as to fully understand what is being done. Implement and run each program in this book yourself.

Download the example code files

You can download the example code files for this book from your account at
`www.packtpub.com`. If you purchased this book elsewhere, you can visit
`www.packtpub.com/support` and register to have the files emailed directly to you.

You can download the code files by following these steps:

1. Log in or register at `www.packtpub.com`.
2. Select the **SUPPORT** tab.
3. Click on **Code Downloads & Errata**.
4. Enter the name of the book in the **Search** box and follow the onscreen instructions.

Once the file is downloaded, please make sure that you unzip or extract the folder using the latest version of:

- WinRAR/7-Zip for Windows
- Zipeg/iZip/UnRarX for Mac
- 7-Zip/PeaZip for Linux

The code bundle for the book is also hosted on GitHub
at `https://github.com/PacktPublishing/Kotlin-Programming-By-Example`. In case there's an update to the code, it will be updated on the existing GitHub repository.

We also have other code bundles from our rich catalog of books and videos available
at `https://github.com/PacktPublishing/`. Check them out!

Download the color images

We also provide a PDF file that has color images of the screenshots/diagrams used in this book. You can download it here: `https://www.packtpub.com/sites/default/files/downloads/KotlinProgrammingByExample_ColorImages.pdf`.

Conventions used

There are a number of text conventions used throughout this book.

`CodeInText`: Indicates code words in text, database table names, folder names, filenames, file extensions, pathnames, dummy URLs, user input, and Twitter handles. Here is an example: "In Java, an `HttpsURLConnection` can be used for secure data transfer over a network."

A block of code is set as follows:

```
release {
    storeFile file("../my-release-key.jks")
    storePassword "password"
    keyAlias "my-alias"
    keyPassword "password"
}
```

When we wish to draw your attention to a particular part of a code block, the relevant lines or items are set in bold:

```
release {
    storeFile file("../my-release-key.jks")
    storePassword "password"
    keyAlias "my-alias"
    keyPassword "password"
}
```

Any command-line input or output is written as follows:

```
./gradlew assembleRelease
```

Bold: Indicates a new term, an important word, or words that you see onscreen. For example, words in menus or dialog boxes appear in the text like this. Here is an example: "Enter the rest of your account details as required and click **COMPLETE REGISTRATION** to finish the account registration process."

Warnings or important notes appear like this.

 Tips and tricks appear like this.

Get in touch

Feedback from our readers is always welcome.

General feedback: Email `feedback@packtpub.com` and mention the book title in the subject of your message. If you have questions about any aspect of this book, please email us at `questions@packtpub.com`.

Errata: Although we have taken every care to ensure the accuracy of our content, mistakes do happen. If you have found a mistake in this book, we would be grateful if you would report this to us. Please visit `www.packtpub.com/submit-errata`, selecting your book, clicking on the Errata Submission Form link, and entering the details.

Piracy: If you come across any illegal copies of our works in any form on the Internet, we would be grateful if you would provide us with the location address or website name. Please contact us at `copyright@packtpub.com` with a link to the material.

If you are interested in becoming an author: If there is a topic that you have expertise in and you are interested in either writing or contributing to a book, please visit `authors.packtpub.com`.

Reviews

Please leave a review. Once you have read and used this book, why not leave a review on the site that you purchased it from? Potential readers can then see and use your unbiased opinion to make purchase decisions, we at Packt can understand what you think about our products, and our authors can see your feedback on their book. Thank you!

For more information about Packt, please visit `packtpub.com`.

The Fundamentals 1

Learning a programming language is a daunting experience for many people, and not one that most individuals generally choose to undertake. As you have chosen to pick up this book, I assume that you have an interest in learning the Kotlin programming language and perhaps even becoming an expert at it someday. As a consequence, permit me to congratulate you on taking a bold step toward learning this language.

Regardless of the problem domain that you may wish to build solutions for, be it application development, networking, or distributed systems, Kotlin is a good choice for the development of systems to achieve the required solutions. In other words, a developer can't go wrong with learning Kotlin. At this point, a proper introduction to the Kotlin language is required.

Kotlin is a strongly-typed, object-oriented language that runs on the **Java Virtual Machine (JVM)** and can be used to develop applications in numerous problem domains. In addition to running on the JVM, Kotlin can be compiled to JavaScript, and as such, is an equally strong choice for developing client-side web applications. Kotlin can also be compiled directly into native binaries that run on systems without a virtual machine via Kotlin/Native. The Kotlin programming language was primarily developed by JetBrains – a company based in Saint Petersburg, Russia. The developers at JetBrains are the current maintainers of the language. Kotlin was named after Kotlin island – an island near Saint Petersburg.

Kotlin was designed to be used to develop industrial-strength software in many domains, but has seen the majority of its use come from the Android ecosystem. At the time of writing, Kotlin is one of the three languages that have been declared by Google as an official language for Android. Kotlin is syntactically similar to Java. As a matter of fact, it was designed to be a better alternative to Java. As a consequence, there are numerous significant advantages to using Kotlin instead of Java in software development.

In this chapter, you will learn the following:

- The installation of Kotlin
- The fundamentals of the Kotlin programming language
- Installing and setting up Android Studio
- Gradle
- The fundamentals of the web

Getting started with Kotlin

In order to develop the Kotlin program, you will first need to install the **Java Runtime Environment** (**JRE**) on your computer. The JRE can be downloaded prepackaged along with a **Java Development Kit** (**JDK**). For the sake of this installation, we will be using the JDK.

The easiest way to install a JDK on a computer is to utilize one of the JDK installers made available by Oracle (the owners of Java). There are different installers available for all major operating systems. Releases of the JDK can be downloaded from
`http://www.oracle.com/technetwork/java/javase/downloads/index.html:`

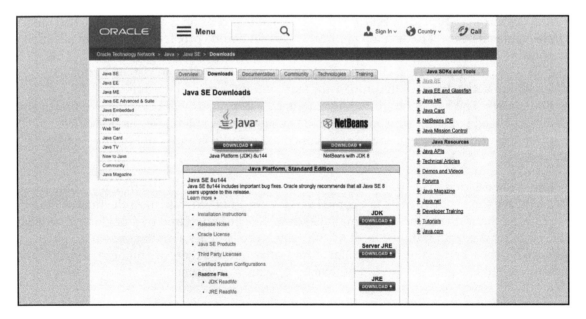

Java SE web page

Clicking on the JDK download button takes you to a web page where you can download the appropriate JDK for your operating system and CPU architecture. Download a JDK suitable for your computer and continue to the next section:

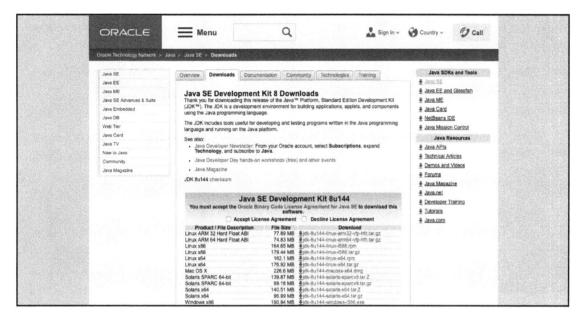

JDK download page

JDK installation

In order to install the JDK on your computer, check out the necessary installation information from the following sections, based on your operating system.

Installation on Windows

The JDK can be installed on Windows in four easy steps:

1. Double-click the downloaded installation file to launch the JDK installer.
2. Click the **Next** button in the welcome window. This action will lead you to a window where you can select the components you want to install. Leave the selection at the default and click **Next**.

3. The following window prompts the selection of the destination folder for the installation. For now, leave this folder as the default (also take note of the location of this folder, as you will need it in a later step). Click **Next**.

4. Follow the instructions in the upcoming windows and click **Next** when necessary. You may be asked for your administrator's password, enter it when necessary. Java will be installed on your computer.

After the JDK installation has concluded, you will need to set the JAVA_HOME environment variable on your computer. To do this:

1. Open your Control Panel.

2. Select **Edit environment variable**.

3. In the window that has opened, click the **New** button. You will be prompted to add a new environment variable.

4. Input JAVA_HOME as the variable name and enter the installation path of the JDK as the variable value.

5. Click OK once to add the environment variable.

Installation on macOS

In order to install the JDK on macOS, perform the following steps:

1. Download your desired JDK .dmg file.

2. Locate the downloaded .dmg file and double-click it.

3. A finder window containing the JDK package icon is opened. Double-click this icon to launch the installer.

4. Click **Continue** on the introduction window.

5. Click **Install** on the installation window that appears.

6. Enter the administrator login and password when required and click **Install Software**.

The JDK will be installed and a confirmation window displayed.

Installation on Linux

Installation of the JDK on Linux is easy and straightforward using `apt-get`:

1. Update the package index of your computer. From your terminal, run:

   ```
   sudo apt-get update
   ```

2. Check whether Java is already installed by running the following:

   ```
   java -version
   ```

3. You'll know Java is installed if the version information for a Java install on your system is printed. If no version is currently installed, run:

   ```
   sudo apt-get install default-jdk
   ```

That's it! The JDK will be installed on your computer.

Compiling Kotlin programs

Now that we have the JDK set up and ready for action, we need to install a means to actually compile and run our Kotlin programs.

Kotlin programs can be either compiled directly with the Kotlin command-line compiler or built and run with the **Integrated Development Environment (IDE)**.

Working with the command-line compiler

The command-line compiler can be installed via Homebrew, SDKMAN!, and MacPorts. Another option for setting up the command-line compiler is by manual installation.

Installing the command-line compiler on macOS

The Kotlin command-line compiler can be installed on macOS in various ways. The two most common methods for its installation on macOS are via Homebrew and MacPorts.

Homebrew

Homebrew is a package manager for the macOS systems. It is used extensively for the installation of packages required for building software projects. To install Homebrew, locate your macOS terminal and run:

```
/usr/bin/ruby -e "$(curl -fsSL
https://raw.githubusercontent.com/Homebrew/install/master/install)"
```

You will have to wait a few seconds for the download and installation of Homebrew. After installation, check to see whether Homebrew is working properly by running the following command in your terminal:

```
brew -v
```

If the current version of Homebrew installed on your computer is printed out in the terminal, Homebrew has been successfully installed on your computer.

After properly installing Homebrew, locate your terminal and execute the following command:

```
brew install kotlin
```

Wait for the installation to finish, after which you are ready to compile Kotlin programs with the command-line compiler.

MacPorts

Similar to HomeBrew, MacPorts is a package manager for macOS. Installing MacPorts is easy. It can be installed on a system by:

1. Installing Xcode and the Xcode command-line tools.
2. Agreeing to the Xcode license. This can be done in the terminal by running `xcodebuild -license`.
3. Installing the required version of MacPorts.

MacPort versions can be downloaded from `https://www.macports.org/install.php`.

Once downloaded, locate your terminal and run `port install kotlin` as the superuser:

```
sudo port install kotlin
```

Installing the command-line compiler on Linux

Linux users can easily install the command-line compiler for Kotlin with SDKMAN!

SDKMAN!

This can be used to install packages on Unix-based systems such as Linux and its various distributions, for example, Fedora and Solaris. SDKMAN! can be installed in three easy steps:

1. Download the software on to your system with `curl`. Locate your terminal and run:

   ```
   curl -s "https://get.sdkman.io" | bash
   ```

2. After you run the preceding command, a set of instructions will come up in your terminal. Follow these instructions to complete the installation. Upon completing the instructions, run:

   ```
   source "$HOME/.sdkman/bin/sdkman-init.sh"
   ```

3. Run the following:

   ```
   sdk version
   ```

   If the version number of SDKMAN! just installed is printed in your terminal window, the installation was successful.

   Now that we have SDKMAN! successfully installed on our system, we can install the command-line compiler by running:

   ```
   sdk install kotlin
   ```

Installing the command-line compiler on Windows

In order to use the Kotlin command-line compilers on Windows:

1. Download a GitHub release of the software from `https://github.com/JetBrains/kotlin/releases/tag/v1.2.30`
2. Locate and unzip the downloaded file

3. Open the extracted `kotlinc\bin` folder
4. Start the command prompt with the folder path

You can now make use of the Kotlin compiler from your command line.

Running your first Kotlin program

Now that we have our command-line compiler set up, let's try it out with a simple Kotlin program. Navigate to your home directory and create a new file named `Hello.kt`. All Kotlin files have a `.kt` extension appended to the end of the filename.

Open the file you just created in a text editor of your choosing and input the following:

```
// The following program prints Hello world to the standard system output.
fun main (args: Array<String>) {
  println("Hello world!")
}
```

Save the changes made to the program file. After the changes have been saved, open your terminal window and input the following command:

```
kotlinc hello.kt -include-runtime -d hello.jar
```

The preceding command compiles your program into an executable, `hello.jar`. The `-include-` runtime flag is used to specify that you want the compiled JAR to be self-contained. By adding this flag to the command, the Kotlin runtime library will be included in your JAR. The `-d` flag specifies that, in this case, we want the output of the compiler to be called.

Now that we have compiled our first Kotlin program, we need to run it—after all, there's no fun in writing programs if they can't be run later on. Open your terminal, if it's not already open, and navigate to the directory where the JAR was saved to (in this case, the home directory). To run the compiled JAR, perform the following:

```
java -jar hello.jar
```

After running the preceding command, you should see `Hello world!` printed on your display. Congratulations, you have just written your first Kotlin program!

Writing scripts with Kotlin

As previously stated, Kotlin can be used to write scripts. Scripts are programs that are written for specific runtime environments for the common purpose of automating the execution of tasks. In Kotlin, scripts have the .kts file extension appended to the file name.

Writing a Kotlin script is similar to writing a Kotlin program. In fact, a script written in Kotlin is exactly like a regular Kotlin program! The only significant difference between a Kotlin script and regular Kotlin program is the absence of a main function.

Create a file in a directory of your choosing and name it NumberSum.kts. Open the file and input the following program:

```
val x: Int = 1
val y: Int = 2
val z: Int = x + y
println(z)
```

As you've most likely guessed, the preceding script will print the sum of 1 and 2 to the standard system output. Save the changes to the file and run the script:

```
kotlinc -script NumberSum.kts
```

 A significant thing to take note of is that a Kotlin script does not need to be compiled.

Using the REPL

REPL is an acronym that stands for **Read–Eval–Print Loop**. An REPL is an interactive shell environment in which programs can be executed with immediate results given. The interactive shell environment can be invoked by running the kotlinc command without any arguments.

 The Kotlin REPL can be started by running kotlinc in your terminal.

If the REPL is successfully started, a welcome message will be printed in your terminal followed by >>> on the next line, alerting us that the REPL is awaiting input. Now you can type in code within the terminal, as you would in any text editor, and get immediate feedback from the REPL. This is demonstrated in the following screenshot:

```
Welcome to Kotlin version 1.1.4-3 (JRE 1.8.0_65-b17)
Type :help for help, :quit for quit
>>> val x: Int = 1
>>> val y: Int = 2
>>> val z: Int = x + y
>>> println(z)
3
>>>
```

Kotlin REPL

In the preceding screenshot, the 1 and 2 integers are assigned to x and y, respectively. The sum of x and y is stored in a new z variable and the value held by z is printed to the display with the `print()` function.

Working with an IDE

Writing programs with the command line has its uses, but in most cases, it is better to use software built specifically for the purpose of empowering developers to write programs. This is especially true in cases where a large project is being worked on.

An IDE is a computer application that hosts a collection of tools and utilities for computer programmers for software development. There are a number of IDEs that can be used for Kotlin development. Out of these IDEs, the one with the most comprehensive set of features for the purpose of developing Kotlin applications is IntelliJ IDEA. As IntelliJ IDEA is built by the creators of Kotlin, there are numerous advantages in using it over other IDEs, such as an unparalleled feature set of tools for writing Kotlin programs, as well as timely updates that cater to the newest advancements and additions to the Kotlin programming language.

Installing IntelliJ IDEA

IntelliJ IDEA can be downloaded for Windows, macOS, and Linux directly from JetBrains' website: `https://www.jetbrains.com/idea/download`. On the web page, you are presented with two available editions for download: a paid Ultimate edition and a free Community edition. The Community edition is sufficient if you wish to run the programs in this chapter. Select the edition you wish to download:

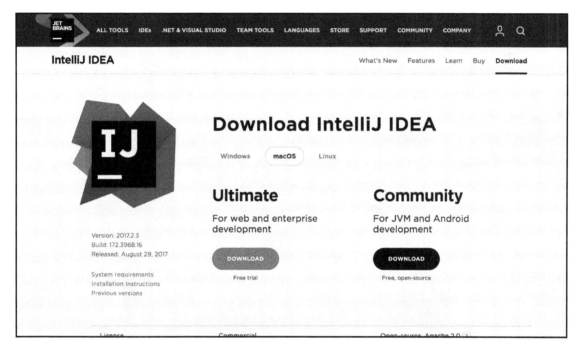

IntelliJ IDEA download page

Once the download is complete, double-click on the downloaded file and install it on your operating system as you would any program.

Setting up a Kotlin project with IntelliJ

The process of setting up a Kotlin project with IntelliJ is straightforward:

1. Start the IntelliJ IDE application.
2. Click **Create New Project**.

3. Select **Java** from the available project options on the left-hand side of the newly opened window.
4. Add Kotlin/JVM as an additional library to the project.
5. Pick a project SDK from the drop-down list in the window.
6. Click **Next**.
7. Select a template if you wish to use one, then continue to the next screen.
8. Provide a project name in the input field provided. Name the project `HelloWorld` for now.
9. Set a project location in the input field.
10. Click **Finish**.

Your project will be created and you will be presented with the IDE window:

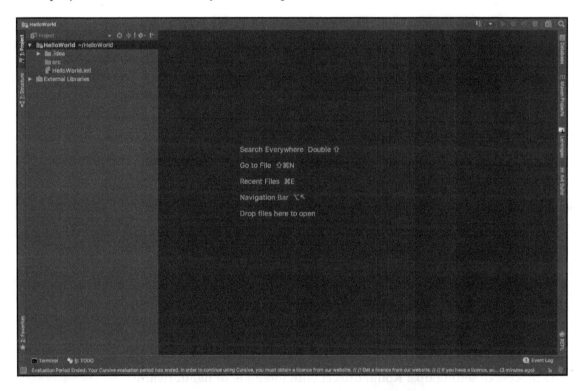

To the left of the window, you will immediately see the project view. This view shows the logical structure of your project files.

Two folders are present. These are:

- `.idea`: This contains IntelliJ's project-specific settings files.
- `src`: This is the source folder of your project. You will place your program files in this folder.

Now that the project is set up, we will write a simple program. Add a file named `hello.kt` to the source folder (right-click the `src` folder, select **New** | **Kotlin File/Class**, and name the file `hello`). Copy and paste the following code into the file:

```
fun main(args: Array<String>) {
  println("Hello world!")
}
```

To run the program, click the Kotlin logo adjacent to the main function and select **Run HelloKt**:

The project will be built and run, after which, `Hello world!` will be printed to the standard system output.

The fundamentals of the Kotlin programming language

Now that we have set up our development environment and our IDE of choice, it is time to explore Kotlin. We will start by diving into the basics of the language and progress into more advanced topics, such as **object-oriented programming** (**OOP**).

Kotlin basics

In this section, we will explore the basics of Kotlin—the building blocks, if you will. We will start by discussing variables.

Variables

A variable is an identifier for a memory location that holds a value. A simpler way to describe a variable is an identifier that holds a value. Consider the following program:

```
fun main(args: Array<String>) {
    var x: Int = 1
}
```

The preceding x is a variable and the value it holds is 1. More specifically, x is an integer variable. The x is referred to as an integer variable because x has been defined to have the Int data type. As such, the x variable can only hold an integer value. To be more accurate, we say that x is an instance of the Int class. At this point, you must be wondering what the words instance and class mean in this context. All will be revealed in due time. For now, let's focus on the topic of variables.

When defining a variable in Kotlin, we make use of the var keyword. This keyword specifies that the variable is mutable in nature. Thus, it can be changed. The data type of the declared variable comes after a semicolon that follows the variable's identifier. It is important to note that the data type of a variable need not be explicitly defined. This is because Kotlin supports type inference—the ability to infer types of objects upon definition. We might as well have written the definition of our x variable as:

```
var x = 1
```

The outcome of the definition would be the same. A semicolon can be added to the end of the line of our variable definition but, similar to languages like JavaScript, it is not required:

```
var x = 1 // I am a variable identified by x and I hold a value of 1
var y = 2 // I am a variable identified by y and I hold a value of 2
var z: Int = x + y // I am a variable identified by z and I hold a value of
3
```

If we don't want the values of our variables to change over the course of the execution of our program, we can do so by making them immutable. Immutable variables are defined with the val keyword, as follows:

```
val x = 200
```

Variable scope

The scope of a variable is the region of a program where the variable is of consequence. In other words, the scope of a variable is the region of a program in which the variable can be used. Kotlin variables have block scope. Therefore, the variables can be utilized in all regions that the block they were defined in covers:

```kotlin
fun main(args: Array<String>) {
  // block A begins
  var a = 10
  var i = 1

  while (i < 10) {
    // block B begins
    val b = a / i
    print(b)
    i++
  }
  print(b) // Error occurs: variable b is out of scope
}
```

In the preceding program, we can directly observe the effects of block scope by taking a look at the two blocks. The definition of a function opened a new block. We have labeled to this block as B in our example. Within A, the a and i variables were declared. As such, the scope of the a and i variables exists within A.

A `while` loop was created within A, and as such, a new B block was opened. Loop declarations mark the beginning of new blocks. Within B, a b value is declared. The b value exists in the B scope and can't be used outside its scope. As such, when we attempt to print the value held by b outside the B block, an error will occur.

One thing worth noting is that the a and i variables can still be utilized within the B block. This is because B exists within the scope of A.

Local variables

These are variables that are local to a scope. The a, i, and b variables in our previous example are all local variables.

Operands and operators

An operator is the part of an instruction that specifies the value to be operated on. An operator also carries out a specific operation on its operands. Examples of operators are +, −, *, /, and %. Operators can be categorized based on the type of operations carried out and the number of operands acted upon by the operator.

Based on the type of operations carried out by the operator, we can classify operators into:

- Relational operators
- Assignment operators
- Logical operators
- Arithmetic operators
- Bitwise operators

Operator type	Examples
Relational operators	>, <, >=, <=, ==
Assignment operators	+=, −=, *=, /=, =
Logical operators	&&, ||, !
Arithmetic operators	+, −, *, /
Bitwise operators	and(bits), or(bits), xor(bits), inv(), shl(bits), shr(bits), ushr(bits)

Based on the number of operands acted upon, we have two main types of operators in Kotlin:

- Unary operators
- Binary operators

Operator type	Description	Examples
Unary operator	Requires only one operand	!, ++, − −
Binary operator	Requires two operands	+, −, *, /, %, &&, ||

Types

The type of a variable, with respect to its value space, is the set of possible values that the variable can possess. In many cases, it is useful to be able to explicitly specify the type of value you want to be held by a variable being declared. This can be done using a data type.

Some important types in Kotlin are:

- Int
- Float
- Double
- Boolean
- String
- Char
- Array

Int

This type represents a 32-bit signed integer. When a variable is declared with this type, the value space of the variable is the set of integers, that is, the variable can only hold integer values. We have seen the use of this type several times in our examples so far. The Int type can hold integer values within the range of -2,147,483,648 to 2,147,483,647.

Float

This type represents a single precision 32-bit floating-point number. When used with a variable, this type specifies that the variable can only hold floating-point values. Its range is approximately ±3.40282347E+38F (6-7 significant decimal digits):

```
var pi: Float = 3.142
```

Double

This type represents a double precision 64-bit floating-point number. Similar to the Float type, this type specifies that the variable being declared holds floating-point values. An important difference between the Double and Float types is that Double can hold numbers across a much larger range without overflow. Its range is approximately ±1.79769313486231570E+308 (15 significant decimal digits):

```
var d: Double = 3.142
```

Boolean

The `true` and `false` logical truth values are represented by the `Boolean` type:

```
var t: Boolean = true
var f: Boolean = false
```

Boolean values are operated upon by the `&&`, `||`, and `!` logical operators:

Operator name	Operator	Description	Operator type
Conjunction	&&	Evaluates to true when two of its operands are `true`, otherwise evaluates to `false`.	Binary
Disjunction	||	Evaluates to true when at least one operand is `true`, otherwise evaluates to `false`.	Binary
Negation	!	Inverts the value of its Boolean operand.	Unary

String

A string is a sequence of characters. In Kotlin, strings are represented by the string class. Strings can be easily written by typing out a sequence of characters and surrounding it with double quotation marks:

```
val martinLutherQuote: String = "Free at last, Free at last, Thank God
almighty we are free at last."
```

Char

This type is used to represent characters. A character is a unit of information that roughly corresponds to a grapheme, or a grapheme-like unit or symbol. In Kotlin, characters are of the `Char` type. Characters in single quotes in Kotlin, such as a, $, %, and &, are all examples of characters:

```
val c: Char = 'i' // I am a character
```

Recall we mentioned earlier that a string is a sequence of characters:

```
var c: Char
val sentence: String = "I am made up of characters."

for (character in sentence) {
  c = character // Value of character assigned to c without error
  println(c)
}
```

Array

An array is a data structure consisting of a set of elements or values with each element possessing at least one index or key. Arrays are very useful in storing collections of elements you wish to utilize later on in a program.

In Kotlin, arrays are created using the `arrayOf()` library method. The values you wish to store in the array are passed in a comma-separated sequence:

```
val names = arrayOf("Tobi", "Tonia", "Timi")
```

Each array value has a unique index that both specifies its position in the array and can be used to retrieve the value later on. The set of indices in an array starts with the index, 0, and progresses with an increment of 1.

The value held in any given index position of an array can be retrieved by either calling the `Array#get()` method or by using the `[]` operation:

```
val numbers = arrayOf(1, 2, 3, 4)
println(numbers[0]) // Prints 1
println(numbers.get(1)) // Prints 2
```

At any point in time, the value at a position of an array can be changed:

```
val numbers = arrayOf(1, 2, 3, 4)
println(numbers[0]) // Prints 1
numbers[0] = 23
println(numbers[0]) // Prints 23
```

You can check the size of an array at any time with its `length` property:

```
val numbers = arrayOf(1, 2, 3, 4)
println(numbers.length) // Prints 4
```

Functions

A function is a block of code that can be defined once and reused any number of times. When writing programs, it is best practice to break up complex programmatic processes into smaller units that perform specific tasks. Doing this has many advantages, some of which are:

- **Improving code readability**: It is much easier to read programs that have been broken down into functional units. This is because the scope of the code to be understood at any given point in time is reduced when functions are utilized. The majority of the time, a programmer needs to only write or adjust a section of a large code base. When functions are utilized, the context of the program that needs to be read to ameliorate program logic is restricted to the body of the function in which the logic is written.
- **Improving the maintainability of a code base**: The use of functions in a code base makes it easy to maintain programs. If a change needs to be made to a particular program feature, many times it is as easy as adjusting a function in which the feature has been created.

Declaring functions

Functions are declared with the `fun` keyword. The following is a simple function definition:

```
fun printSum(a: Int, b: Int) {
  print(a + b)
}
```

The function simply prints the sum of two values that have been passed as arguments to it. Function definitions can be broken down into the following components:

- **A function identifier**: The identifier of a function is the name given to it. An identifier is required to refer to the function if we wish to invoke it later on in a program. In the preceding function declaration, `printSum` is the identifier of the function.
- **A pair of parentheses containing a comma-separated list of the arguments being passed as values to the function**: Values passed to a function are called arguments of the function. All arguments passed to the function must have a type. The type definition of an argument follows a semicolon placed after the argument name.

- **A return type specification**: Return types of functions are specified similarly to the way the types of variables and properties are. The return type specification follows the last parenthesis and is done by writing the type after a semicolon.
- **A block containing the body of the function**.

Observing the preceding function, it may appear that it has no return type. This is not `true`, the function has a return type of `Unit`. A unit return type need not be explicitly specified. The function might as well be declared as follows:

```
fun printSum(a: Int, b: Int): Unit {
  print(a + b)
}
```

An identifier is not required for a function. Functions that do not possess an identifier are called anonymous functions. Anonymous functions are present in Kotlin in the form of lambdas.

Invoking functions

Functions are not executed once they are defined. In order for the code within a function to be executed, the function must be invoked. Functions can be invoked as functions, as methods, and indirectly by utilizing the `invoke()` and `call()` methods. The following shows the direct functional invocation using the function itself:

```
fun repeat(word: String, times: Int) {
  var i = 0

  while (i < times) {
    println(word)
    i++
  }
}

fun main(args: Array<String>) {
  repeat("Hello!", 5)
}
```

Compile and run the preceding code. The word `Hello` is printed on the screen five times. `Hello` was passed as our first value in the function and 5 as our second. As a result of this, the `word` and `times` arguments are set to hold the `Hello` and 5 values in our repeat function. Our `while` loop runs and prints our word as long as `i` is less than the number of times specified. `i++` is used to increase the value of `i` by 1. `i` is increased by one upon each iteration of the loop. The loop stops once `i` is equal to 5. Hence, our word `Hello` will be printed five times. Compiling and running the program will give us the following output:

```
root@vultr:~/kotlin_by_example# java -jar WordRepeat.jar
Hello!
Hello!
Hello!
Hello!
Hello!
root@vultr:~/kotlin_by_example#
```

The other methods of function invocation will be demonstrated over the course of this book.

Return values

A return value—as the name implies—is the value that a method returns. Functions in Kotlin can return values upon execution. The type of the value returned by a function is defined by the function's return type. This is demonstrated in the following code snippet:

```kotlin
fun returnFullName(firstName: String, surname: String): String {
    return "${firstName} ${surname}"
}

fun main(args: Array<String>) {
    val fullName: String = returnFullName("James", "Cameron")
    println(fullName) // prints: James Cameron
}
```

In the preceding code, the `returnFullName` function takes two distinct strings as its input parameters and returns a string value when called. The return type has been defined in the function header. The string returned is created via string templates:

```kotlin
"${firstName} ${surname}"
```

The values for first name and last name are interpolated into the string of characters.

The function naming convention

The conventions for naming functions in Kotlin are similar to that of Java. When naming methods, camel case is utilized. In camel case, names are written such that each word in the name begins with a capital letter, with no intervening spaces or punctuation:

```
//Good function name
fun sayHello() {
  println("Hello")
}

//Bad function name
fun say_hello() {
  println("Hello")
}
```

Comments

When writing code, you may need to jot down important information pertaining to the code being written. This is done through the use of comments. There are three types of comments in Kotlin:

- Single-line comments
- Multiline comments
- Doc comments

Single-line comments

As the name implies, these comments span a single line. Single-line comments are started with two backslashes (//). Upon compilation of your program, all characters coming after these slashes are ignored. Consider the following code:

```
val b: Int = 957 // This is a single line comment
// println(b)
```

The value held by b is never printed to the console because the function that performs the printing operation has been commented out.

Multiline comments

Multiline comments span multiple lines. They are started with a backslash followed by an asterisk (/*) and ended by an asterisk followed by a backslash (*/):

```
/*
 * I am a multiline comment.
 * Everything within me is commented out.
 */
```

Doc comments

This type of comment is similar to a multiline comment. The major difference is that it is used to document code within a program. A doc comment starts with a backslash followed by two asterisk characters (/**) and ends with an asterisk followed by a backslash (*/):

```
/**
 * Adds an [item] to the queue.
 * @return the new size of the queue.
 */
fun enqueue(item: Object): Int { ... }
```

Controlling program flow

When writing programs, a scenario that often occurs is one in which we want to control how our program executes. This is necessary if we want to write programs that can make decisions based on conditions and program state. Kotlin possesses a number of structures for doing this, which will be familiar to people who have worked with programming languages in the past, such as if, while, and for constructs. There are also others that may not be familiar to individuals, such as the when construct. In this section, we will take a look at the structures at our disposal for controlling the flow of our program.

Conditional expressions

Conditional expressions are used for branching program flow. They execute or skip program statements based on the outcome of a conditional test. Conditional statements are the decision-making points of a program.

Kotlin has two main structures for handling branching. These are if expressions and when expressions.

The if expression

The `if` expression is used to make a logical decision based on the fulfillment of a condition. We make use of the `if` keyword to write `if` expressions:

```
val a = 1

if (a == 1) {
 print("a is one")
}
```

The preceding `if` expression tests whether the `a == 1` (read: a is equal to 1) condition holds `true`. If the condition is `true`, the `a is one` string is printed on the screen, otherwise nothing is printed.

An `if` expressions often has one or more accompanying `else` or `else if` keywords. These accompanying keywords can be used to further control the flow of a program. Take the following `if` expression for example:

```
val a = 4
if (a == 1) {
   print("a is equal to one.")
} else if (a == 2) {
    print("a is equal to two.")
} else {
    print("a is neither one nor two.")
}
```

The preceding expression first tests whether a is equal to 1. This test evaluates to `false` so, the following condition has been tested. Surely a is not equal to 2. Hence the second condition evaluates to `false`. As a result of all previous conditions evaluating to `false`, the final statement is executed. Hence `a is neither one nor two.` is printed on the screen.

The when expression

The `when` expression is another means of controlling program flow. Let's observe how it works with a simple example:

```
fun printEvenNumbers(numbers: Array<Int>) {
  numbers.forEach {
    when (it % 2) {
      0 -> println(it)
    }
  }
}
```

```
fun main (args: Array<String>) {
    val numberList: Array<Int> = arrayOf(1, 2, 3, 4, 5, 6)
    printEvenNumbers(numberList)
}
```

The preceding `printEvenSum` function takes an integer array as its only argument. We will cover arrays later on in this chapter, but for now think of them as a sequential collection of values existing in a value space. In this case, the array passed contains values that exist in the value space of integers. Each element of the array is iterated upon using the `forEach` method and each number is tested in the `when` expression.

Here, the `it` refers to the current value being iterated upon by the `forEach` method. The `%` operator is a binary operator that acts on two operands. It divides the first operand by the second and returns the remainder of the division. Thus, the `when` expression tests if/when the current value iterated upon (the value held within `it`) is divided by 2 and has a remainder of 0. If it does, the value is even and hence the value is printed.

To observe how the program works, copy and paste the preceding code into a file, then compile and run the program:

```
root@vultr:~/kotlin_by_example# java -jar EvenPrint.jar
2
4
6
root@vultr:~/kotlin_by_example#
```

The Elvis operator

The Elvis operator is a terse structure that is present in Kotlin. It takes the following form:

```
(expression) ?: value2
```

Its usage in a Kotlin program is demonstrated in the following code block:

```
val nullName: String? = null
val firstName = nullName ?: "John"
```

If the value held by `nullName` is not null, the Elvis operator returns it, otherwise the `"John"` string is returned. Thus, `firstName` is assigned the value returned by the Elvis operator.

Loops

Looping statements are used to ensure that a collection of statements within a block of code repeats in execution. That is, a loop ensures that a number of statements within a program executes for a number of times. The looping constructs provided by Kotlin are the `for` loop, the `while` loop, and the `do...while` loop.

The for loops

The `for` loop in Kotlin iterates over any object that provides an iterator. It is similar to the `for..in` loop in Ruby. The loop has this syntax:

```
for (obj in collection) { ... }
```

The block in the `for` loop is not necessary if only a single statement exists in the loop. A collection is a type of structure that provides an iterator. Consider the following program:

```
val numSet = arrayOf(1, 563, 23)

for (number in numSet) {
    println(number)
}
```

Each value in the `numSet` array is iterated upon by the loop and assigned to the variable number. The number is then printed to the standard system output.

 Every element of an array has an index. An index is the position an element holds within an array. The set of indices of an array in Kotlin starts from zero.

If instead of printing the numeric values of the number iterated upon, we wish to print the indices of each number, we can do that as follows:

```
for (index in numSet.indices) {
    println(index)
}
```

You can specify a type for your iterator variable as well:

```
for (number: Int in numSet) {
  println(number)
}
```

The while loops

A `while` loop executes instructions within a block as long as a specified condition is met. The `while` loops are created using the `while` keyword. It takes the following form:

```
while (condition) { ... }
```

As in the case of the `for` loop, the block is optional in the case where only one sentence is within the scope of the loop. In a `while` loop, the statements in the block execute repeatedly while the condition specified still holds. Consider the following code:

```
val names = arrayOf("Jeffrey", "William", "Golding", "Segun", "Bob")
var i = 0

while (!names[i].equals("Segun")) {
  println("I am not Segun.")
  i++
}
```

In the preceding program, the block of code within the `while` loop executes and prints I am not Segun until the name Segun is encountered. Once Segun is encountered, the loop terminates and nothing else is printed out, as shown in the following screenshot:

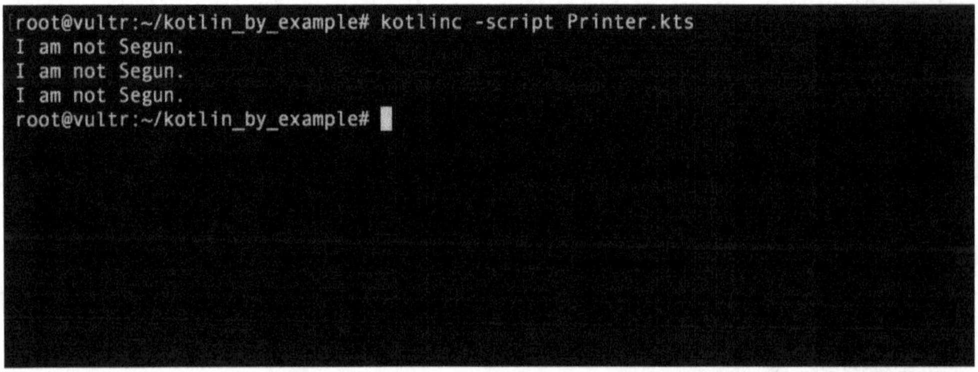

```
root@vultr:~/kotlin_by_example# kotlinc -script Printer.kts
I am not Segun.
I am not Segun.
I am not Segun.
root@vultr:~/kotlin_by_example#
```

The break and continue keywords

Often when declaring loops, there is a need to either break out of the loop if a condition fulfills, or start the next iteration at any point in time within the loop. This can be done with the break and continue keywords. Let's take an example to explain this further. Open a new Kotlin script file and copy the following code into it:

```kotlin
data class Student(val name: String, val age: Int, val school: String)

val prospectiveStudents: ArrayList<Student> = ArrayList()
val admittedStudents: ArrayList<Student> = ArrayList()

prospectiveStudents.add(Student("Daniel Martinez", 12, "Hogwarts"))
prospectiveStudents.add(Student("Jane Systrom", 22, "Harvard"))
prospectiveStudents.add(Student("Matthew Johnson", 22, "University of
Maryland"))
prospectiveStudents.add(Student("Jide Sowade", 18, "University of Ibadan"))
prospectiveStudents.add(Student("Tom Hanks", 25, "Howard University"))

for (student in prospectiveStudents) {
  if (student.age < 16) {
    continue
  }
  admittedStudents.add(student)

  if (admittedStudents.size >= 3) {
    break
  }
}

println(admittedStudents)
```

The preceding program is simplistic software for selecting admitted students out of a list of prospective students. We create a data class at the start of our program to model the data of each student, then two array lists are created. One array list holds the information of the prospective students, those that have applied for admission, and the other list holds the information of the students that have been admitted.

The next five lines of code add prospective students to the prospective student list. We then declare a loop that iterates over all students present in the prospective student list. If the age of the current student in the loop is less than 16 years old, the loop skips to the next iteration. This models the scenario in where a student is too young to be admitted (thus not added to the admitted students list).

If the student is 16 or older, the student is added to the admitted list. An `if` expression is then used to check whether the number of admitted students is greater than or equal to three. If the condition is true, the program breaks out of the loop and no further iterations are done. The last line of the program prints out the students present in the list.

Run the program to see the output:

```
root@vultr:~/kotlin_by_example# kotlinc -script StudentAdmitter.kts
[Student(name=Jane Systrom, age=22, school=Harvard), Student(name=Matthew Johns
on, age=22, school=University of Maryland), Student(name=Jide Sowade, age=18, s
chool=University of Ibadan)]
root@vultr:~/kotlin_by_example#
```

The do...while loops

The `do...while` loop is similar to the `while` loop, with the exception that in the conditional test for the reiteration of the loop, it is carried out after the first iteration. It takes the following form:

```
do {
...
} while (condition)
```

The statements within the block are executed while the condition tested holds `true`:

```
var i = 0

do {
  println("I'm in here!")
  i++
} while (i < 10)

println("I'm out here!")
```

Nullable values: The `NullPointerException` is one thing that individuals who have first-hand experience writing Java code are certain to have encountered. The Kotlin type system is null-safe—it attempts to eliminate the occurrence of null references within code. As a result, Kotlin possesses nullable types and non-nullable types (types that can hold a null value and those that can't).

To properly explain the `NullPointerException`, we will consider the following Java program:

```java
class NullPointerExample {

  public static void main(String[] args) {
    String name = "James Gates";
    System.out.println(name.length()); // Prints 11

    name = null; // assigning a value of null to name
    System.out.println(name.length()); // throws NullPointerException
  }
}
```

The preceding program performs the simple task of printing the length of a string variable to the standard system output. There is only one problem with our program. When we compile and run it, it throws a null pointer exception and terminates midway through execution, as we can see the following screenshot:

```
root@vultr:~/kotlin_by_example# java NullPointerExample
11
Exception in thread "main" java.lang.NullPointerException
        at NullPointerExample.main(NullPointerExample.java:7)
root@vultr:~/kotlin_by_example#
```

Can you spot the cause of the `NullPointerException`? The exception arises as a result of the `String#length` method being used on a null reference. As such, the program stops executing and throws the exception. Clearly, this is not something we want to occur in our programs.

We can prevent this in Kotlin by preventing the assignment of a null value to the name object:

```kotlin
var name: String = "James Gates"
println(name.length)

name = null // null value assignment not permitted
println(name.length)
```

As can be seen in the following screenshot, Kotlin's type system detects that a null value has been inappropriately assigned to the `name` variable and swiftly alerts the programmer of this blunder so it can be corrected:

```
root@vultr:~/kotlin_by_example# kotlinc -script NullSafety.kts
NullSafety.kts:4:8: error: null can not be a value of a non-null type String
name = null
       ^
root@vultr:~/kotlin_by_example#
```

At this point, you may be wondering what happens if a scenario arises in which the programmer intends to permit the passing of null values. In that scenario, the programmer simply declares the value as nullable by appending ? to the type of the variable:

```
var name: String? = "James"
println(name.length)

name = null // null value assignment permitted
println(name.length)
```

Regardless of the fact that we have declared the variable name to be nullable, we'll still get an error upon running the program. This is because we must access the length property of the variable in a safe way. This can be done with ? . :

```
var name: String? = "James"
println(name?.length)

name = null // null value assignment permitted
println(name?.length)
```

Now that we have used the ?. safe operator, the program will run as intended. Instead of a null pointer exception being thrown, the type system recognizes that a null pointer has been referenced and prevents the invocation of length() on the null object. The following screenshot shows the type-safe output:

```
root@vultr:~/kotlin_by_example# kotlinc -script NullSafety.kts
11
null
root@vultr:~/kotlin_by_example#
```

An alternative to using the ?. safe operator would be to use the !! operator. The !! operator allows the program to continue execution and throws a KotlinNullPointerException once a function invocation is attempted on a null reference.

We can see the effects by replacing ?. with !!. in our written program. The following screenshot shows the output of the program when run. A KotlinNullPointerException is thrown as a result of the utilization of the !! operator:

```
root@vultr:~/kotlin_by_example# kotlinc -script NullSafety.kts
11
kotlin.KotlinNullPointerException
        at NullSafety.<init>(NullSafety.kts:5)
root@vultr:~/kotlin_by_example#
```

Packages

A package is a logical grouping of related classes, interfaces, enumerations, annotations, and functions. As source files grow larger, it is necessary to group these files into meaningful and distinct collections for various reasons, such as to enhance the maintainability of applications, name conflict prevention, and access control.

A package is created with the `package` keyword followed by a package name:

```
package foo
```

There can be only one package statement per program file. If a package for a program file is not specified, the contents of the file are placed into the default package.

The import keyword

Often, classes and types need to make use of other classes and types existing outside the package in which they are declared. This can be done by importing package resources. If two classes belong in the same package, no import is necessary:

```
package animals
data class Buffalo(val mass: Int, val maxSpeed: Int, var isDead: Boolean =
false)
```

In the following code snippet, the `Buffalo` class does not need to be imported into the program because it exists in the same package (`animals`) as the `Lion` class:

```
package animals
class Lion(val mass: Int, val maxSpeed: Int) {

  fun kill(animal: Buffalo) { // Buffalo type used with our import
    if (!animal.isDead) {
      println("Lion attacking animal.")
      animal.isDead = true
      println("Lion kill successful.")
    }
  }
}
```

In order to import classes, functions, interfaces, and types in separate packages, we use the `import` keyword followed by the package name. For example, the following `main` function exists in the default package. As such, if we want to make use of the `Lion` and `Buffalo` classes in the `main` function, we must import it with the `import` keyword. Consider the following code:

```
import animals.Buffalo
import animals.Lion

fun main(args: Array<String>) {
  val lion = Lion(190, 80)
  val buffalo = Buffalo(620, 60)
  println("Buffalo is dead: ${buffalo.isDead}")
  lion.kill(buffalo)
  println("Buffalo is dead: ${buffalo.isDead}")
}
```

Object-oriented programming concepts

Up till now, we have made used of classes in a number of examples but have not explored the concept in depth. This section will introduce you to the basics of classes as well as other object-oriented constructs available in Kotlin.

Introduction

In the beginning of high-level programming languages, programs were written procedurally. The programming languages available were mainly procedural in nature. A procedural programming language is a language that utilizes a series of structured, well-defined steps to compose programs.

As the software industry grew larger and programs grew bulkier, it became necessary to devise a better approach to designing software. This led to the advent of object-oriented programming languages.

The object-oriented programming model is a model organized around objects and data rather than actions and sequential logic. In object-oriented programming, objects, classes, and interfaces are composed, extended, and inherited toward the goal of building industrial-strength software.

A **class** is a modifiable and extensible program template for the creation of objects and the maintenance of state through the use of variables, constants, and properties. A class possesses characteristics and behaviors. Characteristics are exposed as variables and behaviors are implemented in the form of methods. Methods are functions that are specific to a class or a collection of classes. Classes have the ability to inherit characteristics and behaviors from other classes. This ability is called **inheritance**.

Kotlin is a fully object-oriented programming language and hence supports all features of object-oriented programming. In Kotlin, similar to Java and Ruby, only single inheritance is permitted. Some languages, such as C++, support multiple inheritance. A downside to multiple inheritance is that it brings up management issues, such as same name collisions. A class inheriting from another class is referred to as a subclass and the class it inherits from is its superclass.

An **interface** is a structure that enforces certain characteristics and behaviors in classes. Behavioral enforcements via interfaces can be done by implementing an interface in a class. Similar to classes, an interface can extend another interface.

An **object** is an instance of a class that may possesses its own unique state.

Working with classes

A class is declared using the `class` keyword followed by the class name:

```
class Person
```

As in the preceding example, a class in Kotlin need not have a body. Though this is charming, almost all the time you will want your class to have characteristics and behaviors placed within a body. This can be done with the use of opening and closing brackets:

```
class HelloPrinter {
  fun printHello() {
    println("Hello!")
  }
}
```

In the preceding code snippet, we have a class named `HelloPrinter` with a single function declared in it. A function that is declared within a class is called a method. Methods can also be referred to as behaviors. Once a method is declared, it can be used by all instances of the class.

Creating objects

Declaring an instance of a class—or an object—is similar to declaring a variable. We can create an instance of the `HelloPrinter` class, as shown in the following code:

```
val printer = HelloPrinter()
```

The preceding `printer` is an instance of the `HelloPrinter` class. The opening and closing brackets after the `HelloPrinter` class name are used to invoke the primary constructor of the `HelloPrinter` class. A constructor is similar to a function. A constructor is a special function that initializes an object of a type.

The function declared within the `HelloPrinter` class can be invoked directly by the `printer` object at any point in time:

```
printer.printHello() // Prints hello
```

Occasionally, you may require a function to be directly invokable with the class without needing to create an object. This can be done with the use of a companion object.

Companion objects

Companion objects are declared within a class by utilizing the `companion` and `object` keywords. You can use functions that are static within a companion object:

```
class Printer {
  companion object DocumentPrinter {
    fun printDocument() = println("Document printing successful.")
  }
}

fun main(args: Array<String>) {
  Printer.printDocument() // printDocument() invoked via companion object
  Printer.Companion.printDocument() // also invokes printDocument() via
                                    // a companion object
}
```

Sometimes, you may want to give an identifier to a companion object. This can be done by placing the name after the `object` keyword. Consider the following example:

```
class Printer {
  companion object DocumentPrinter { // Companion object identified by
  DocumentPrinter
    fun printDocument() = println("Document printing successful.")
  }
```

```
}

fun main(args: Array<String>) {
    Printer.DocumentPrinter.printDocument() // printDocument() invoked via
                                            // a named companion object
}
```

Properties

Classes can have properties that may be declared using the `var` and `val` keywords. For example, in the following code snippet, the `Person` class has three properties, `age`, `firstName`, and `surname`:

```
class Person {    var age = 0
    var firstName = ""
    var surname = ""
}
```

Properties can be accessed through an instance of the class holding the property. This is done by appending a single dot character (`.`) to an instance identifier followed by the property name. For example, in the following code snippet, an instance—named `person`—of the `Person` class is created and its `firstName`, `surname`, and `age` properties are assigned by accessing the properties:

```
val person = Person()
person.firstName = "Raven"
person.surname = "Spacey"
person.age = 35
```

Advantages of Kotlin

As previously discussed, Kotlin was designed to be a better Java, and as such, there are a number of advantages to using Kotlin over Java:

- **Null safety**: One common occurrence in Java programs is the throwing of `NullPointerException`. Kotlin alleviates this issue by providing a null-safe type system.
- **Presence of extension functions**: Functions can easily be added to classes defined in program files to extend their functionality in various ways. This can be done with extension functions in Kotlin.

- **Singletons**: It is easy to implement the singleton pattern in Kotlin programs. The implementation of a singleton in Java takes considerably more effort than when it is done with Kotlin.
- **Data classes**: When writing programs, it is a common scenario to have to create a class for the sole purpose of holding data in variables. This often leads to the writing of many lines of code for such a mundane task. Data classes in Kotlin make it extremely easy to create such classes that hold data with a single line of code.
- **Function types**: Unlike Java, Kotlin has function types. This enables functions to accept other functions as parameters and the definition of functions that return functions.

Developing Android applications with Kotlin

We have taken a concise look at some of the features Kotlin has put at our disposal for developing powerful applications. Over the course of this book, we will explore how these features can be utilized in Android application development - an area in which Kotlin shines.

Before getting started on this journey, we must set up our systems for the task at hand. A major necessity for developing Android applications is a suitable IDE - it is not a requirement but it makes the development process easier. Many IDE choices exist for Android developers. The most popular are:

- Android Studio
- Eclipse
- IntelliJ IDE

Android Studio is by far the most powerful of the IDEs available with respect to Android development. As a consequence, we will be utilizing this IDE in all Android-related chapters in this book.

Setting up Android Studio

At the time of writing, the version of Android Studio that comes bundled with full Kotlin support is Android Studio 3.0. The canary version of this software can be downloaded from `https://developer.android.com/studio/preview/index.html`. Once downloaded, open the downloaded package or executable and follow the installation instructions. A setup wizard exists to guide you through the IDE setup procedure:

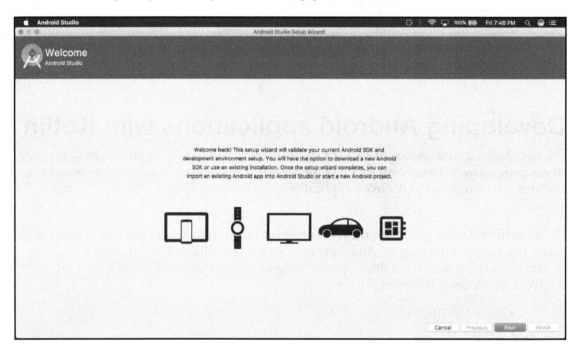

Continuing to the next setup screen will prompt you to choose which type of Android Studio setup you'd like:

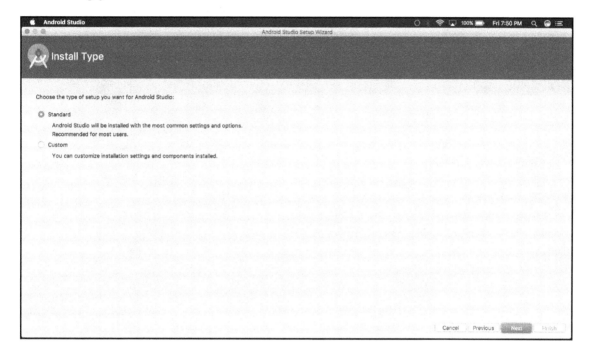

Select the **Standard** setup and continue to the next screen. Click **Finish** on the **Verify Settings** screen. Android Studio will now download the components required for your setup. You will need to wait a few minutes for the required components to download:

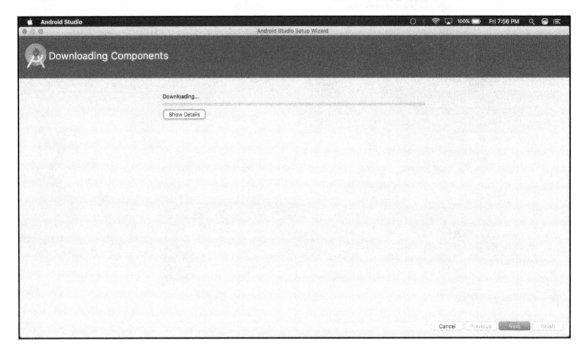

Click **Finish** once the component download has completed. You will be taken to the Android Studio landing screen. You are now ready to use Android Studio:

Building your first Android application

Without further ado, let's explore how to create a simple Android application with Android Studio. We will be building the *HelloApp*. The *HelloApp* is an app that displays `Hello world!` on the screen upon the click of a button.

On the Android Studio landing screen, click **Start a new Android Studio project**. You will be taken to a screen where you will specify some details that concern the app you are about to build, such as the name of the application, your company domain, and the location of the project.

Type in `HelloApp` as the application name and enter a company domain. If you do not have a company domain name, fill in any valid domain name in the company domain input box – as this is a trivial project, a legitimate domain name is not required. Specify the location in which you want to save this project and tick the checkbox for the inclusion of Kotlin support.

After filling in the required parameters, continue to the next screen:

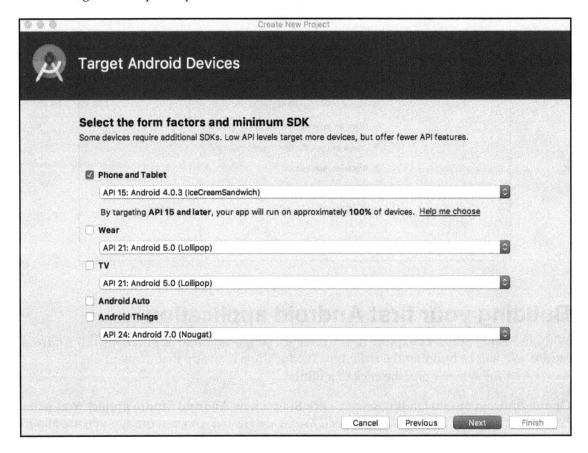

Here, we are required to specify our target devices. We are building this application to run on smartphones specifically, hence tick the **Phone and Tablet** checkbox if it's not already ticked. You will notice an options menu next to each device option. This dropdown is used to specify the target API level for the project being created. An API level is an integer that uniquely identifies the framework API division offered by a version of the Android platform. Select API level 15 if not already selected and continue to the next screen:

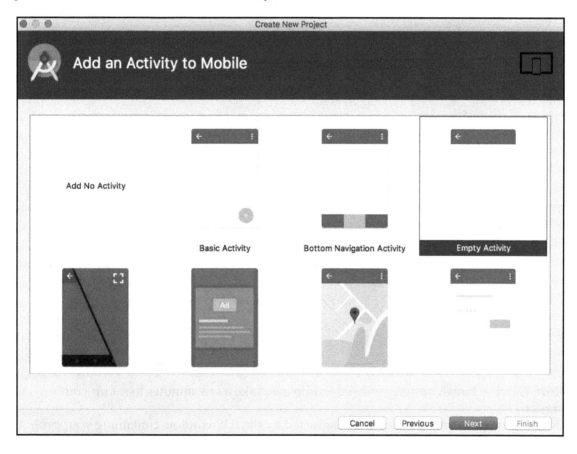

On the next screen, we are required to select an activity to add to our application. An activity is a single screen with a unique user interface—similar to a window. We will discuss activities in more depth in Chapter 2, *Building an Android Application – Tetris*. For now, select the empty activity and continue to the next screen.

Now, we need to configure the activity that we just specified should be created. Name the activity `HelloActivity` and ensure the **Generate Layout File** and **Backwards Compatibility** checkboxes are ticked:

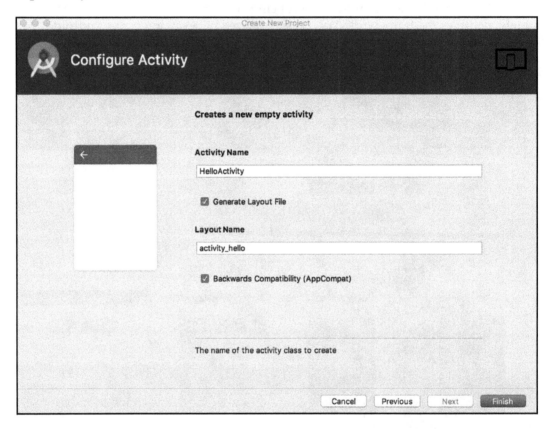

Now, click the **Finish** button. Android Studio may take a few minutes to set up your project.
Once the setup is complete, you will be greeted by the IDE window containing your project files.

 Errors pertaining to the absence of required project components may be encountered at any point during project development. Missing components can be downloaded from the SDK manager.

Make sure that the project window of the IDE is open (on the navigation bar, select **View** | **Tool Windows** | **Project**) and the Android view is currently selected from the drop-down list at the top of the **Project** window. You will see the following files at the left-hand side of the window:

- app | java | com.mydomain.helloapp | HelloActivity.java: This is the main activity of your application. An instance of this activity is launched by the system when you build and run your application:
- app | res | layout | activity_hello.xml: The user interface for HelloActivity is defined within this XML file. It contains a TextView element placed within the ViewGroup of a ConstraintLayout. The text of the TextView has been set to Hello World!
- app | manifests | AndroidManifest.xml: The AndroidManifest file is used to describe the fundamental characteristics of your application. In addition, this is the file in which your application's components are defined.
- Gradle Scripts | build.gradle: Two build.gradle files will be present in your project. The first build.gradle file is for the project and the second is for the app module. You will most frequently work with the module's build.gradle file for the configuration of the compilation procedure of Gradle tools and the building of your app.

 Gradle is an open source build automation system used for the declaration of project configurations. In Android, Gradle is utilized as a build tool with the goal of building packages and managing application dependencies.

Creating a user interface

A **user interface** (**UI**) is the primary means by which a user interacts with an application. The user interfaces of Android applications are made by the creation and manipulation of layout files. Layout files are XML files that exist in app | res | layout.

To create the layout for the *HelloApp*, we are going to do three things:

1. Add a LinearLayout to our layout file
2. Place the TextView within the LinearLayout and remove the android:text attribute it possesses
3. Add a button to the LinearLayout

Open the `activity_hello.xml` file if it's not already opened. You will be presented with the layout editor. If the editor is in the `Design` view, change it to its `Text` view by toggling the option at the bottom of the layout editor. Now, your layout editor should look similar to that of the following screenshot:

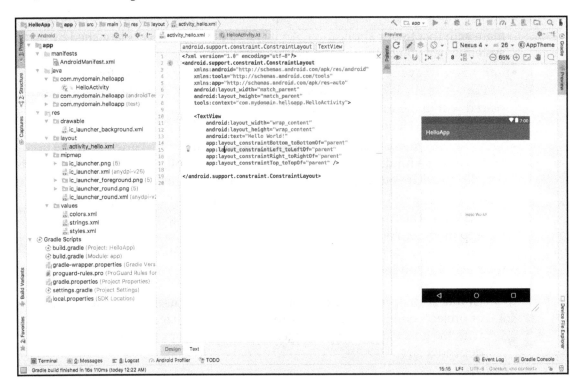

A `LinearLayout` is a `ViewGroup` that arranges child views in either a horizontal or vertical manner within a single column. Copy the code snippet of our required `LinearLayout` from the following block and paste it within the `ConstraintLayout` preceding the `TextView`:

```
<LinearLayout
        android:id="@+id/ll_component_container"
        android:layout_width="match_parent"
        android:layout_height="match_parent"
        android:orientation="vertical"
        android:gravity="center">
</LinearLayout>
```

Now, copy and paste the `TextView` present in the `activity_hello.xml` file into the body of the `LinearLayout` element and remove the `android:text` attribute:

```
<LinearLayout
        android:id="@+id/ll_component_container"
        android:layout_width="match_parent"
        android:layout_height="match_parent"
        android:orientation="vertical"
        android:gravity="center">
    <TextView
        android:id="@+id/tv_greeting"
        android:layout_width="wrap_content"
        android:layout_height="wrap_content"
        android:textSize="50sp" />
</LinearLayout>
```

Lastly, we need to add a button element to our layout file. This element will be a child of our `LinearLayout`. To create a button, we use the `Button` element:

```
<LinearLayout
        android:id="@+id/ll_component_container"
        android:layout_width="match_parent"
        android:layout_height="match_parent"
        android:orientation="vertical"
        android:gravity="center">
    <TextView
        android:id="@+id/tv_greeting"
        android:layout_width="wrap_content"
        android:layout_height="wrap_content"
        android:textSize="50sp" />
    <Button
        android:id="@+id/btn_click_me"
        android:layout_width="wrap_content"
        android:layout_height="wrap_content"
        android:layout_marginTop="16dp"
        android:text="Click me!"/>
</LinearLayout>
```

Toggle to the layout editor's design view to see how the changes we have made thus far translate when rendered on the user interface:

Now we have our layout, but there's a problem. Our **CLICK ME!** button does not actually do anything when clicked. We are going to fix that by adding a listener for click events to the button. Locate and open the HelloActivity.java file and edit the function to add the logic for the **CLICK ME!** button's click event as well as the required package imports, as shown in the following code:

```
package com.mydomain.helloapp
import android.support.v7.app.AppCompatActivity
import android.os.Bundle
import android.text.TextUtils
import android.widget.Button
import android.widget.TextView
import android.widget.Toast

class HelloActivity : AppCompatActivity() {

  override fun onCreate(savedInstanceState: Bundle?) {
    super.onCreate(savedInstanceState)
    setContentView(R.layout.activity_hello)
    val tvGreeting = findViewById<TextView>(R.id.tv_greeting)
    val btnClickMe = findViewById<Button>(R.id.btn_click_me)
```

```
btnClickMe.setOnClickListener {
  if (TextUtils.isEmpty(tvGreeting.text)) {
    tvGreeting.text = "Hello World!"
  } else {
    Toast.makeText(this, "I have been clicked!",
                 Toast.LENGTH_LONG).show()
  }
}
}
}
```

In the preceding code snippet, we have added references to the `TextView` and `Button` elements present in our `activity_hello` layout file by utilizing the `findViewById` function. The `findViewById` function can be used to get references to layout elements that are within the currently-set content view. The second line of the `onCreate` function has set the content view of `HelloActivity` to the `activity_hello.xml` layout.

Next to the `findViewById` function identifier, we have the `TextView` type written between two angular brackets. This is called a function generic. It is being used to enforce that the resource ID being passed to the `findViewById` belongs to a `TextView` element.

After adding our reference objects, we set an `onClickListener` to `btnClickMe`. Listeners are used to listen for the occurrence of events within an application. In order to perform an action upon the click of an element, we pass a lambda containing the action to be performed to the element's `setOnClickListener` method.

When `btnClickMe` is clicked, `tvGreeting` is checked to see whether it has been set to contain any text. If no text has been set to the `TextView`, then its text is set to `Hello World!`, otherwise a toast is displayed with the `I have been clicked!` text.

Running the application

In order to run the application, click the Run 'app' (^R) button at the top-right side of the IDE window and select a deployment target. The *HelloApp* will be built, installed, and launched on the deployment target:

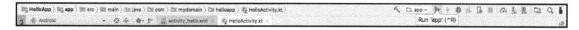

You may use one of the available prepackaged virtual devices or create a custom virtual device to use as the deployment target. You may also decide to connect a physical Android device to your computer via USB and select it as your target. The choice is up to you. After selecting a deployment device, click **OK** to build and run the application.

Upon launching the application, our created layout is rendered:

When **CLICK ME!** is clicked, **Hello World!** is shown to the user:

Subsequent clicks of the **CLICK ME!** button display a toast message with the text **I have been clicked!**:

Fundamentals of the web

Most applications communicate with a server in one way or another. It is imperative that you understand a number of concepts related to the web before continuing in this book. This section explains those concepts concisely.

What is the web?

The web is a complex system of interconnected systems possessing the ability to communicate with other systems existing on a common network via one or more protocols. A protocol is an official, well-defined system of rules governing the exchange of information between devices.

Hypertext Transfer Protocol

All communications over the web are made in accordance with a protocol. A particularly important protocol for fostering communication between systems is the **Hypertext Transfer Protocol** (**HTTP**).

Billions of images, videos, text files, documents, and other files are transferred across the internet on a daily basis. These files are all transferred through HTTP. HTTP is an application protocol for distributed and hypermedia information systems. It can be said to be a foundational component for communication across the internet. A major benefit of using HTTP for data transfer across systems is that it is highly reliable. This is as a result of its utilization of reliable protocols, such as the **Transmission Control Protocol** (**TCP**) and **Internet Protocol** (**IP**).

Clients and servers

A web client is any application that communicates with a web server utilizing HTTP. A web server is a computer that provides – or serves – web resources to web clients. A web resource is anything that provides web content. A web resource can be a media file, an HTML document, a gateway, and so on. Clients need web content for various purposes, such as information rendering and data manipulation.

Clients and servers communicate with each other via HTTP. One major reason for the utilization of HTTP is the fact that it is extremely reliable in data transmission. The use of HTTP ensures data loss does not occur in a request-response cycle.

HTTP requests and responses

An HTTP request – as the name replies – is a solicitation for a web resource sent by a web client to a server over HTTP. An HTTP response is a reply – sent by a server – to a request in an HTTP transaction:

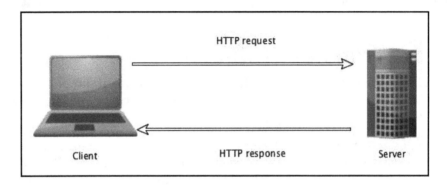

HTTP methods

HTTP supports a number of request methods. These methods can also be referred to as commands. HTTP methods specify the type of action to be performed by the server. Some common HTTP methods are tabularized as follows:

HTTP method	Description
GET	Retrieves a named resource present on the client.
POST	Sends data from a client to a server.
DELETE	Deletes a named resource residing on a server.
PUT	Store data collected by the client in a named resource residing on the server.
OPTIONS	Returns HTTP methods that the server supports.
HEAD	Retrieves HTTP headers with no content.

Summary

In this chapter, we were introduced to Kotlin and explored its fundamentals. In the process, we learned how to install Kotlin on a computer, what an IDE is, the IDEs at your disposal for writing programs in Kotlin, how to write and run Kotlin scripts, and how to use the REPL. In addition, we learned how to work with both IntelliJ IDEA and Android Studio, after which we implemented a simple Android application. Lastly, we had a look at the fundamental concepts related to the web.

In the next chapter, we will get more familiar with writing Kotlin programs by creating an Android application. We will take a look at the Android application architecture and the important components of an Android application, as well as many more topics.

Building an Android Application – Tetris 2

In the previous chapter, we took a concise look at crucial topics pertaining to the core Kotlin language. These topics took us through the fundamentals of Kotlin as well as the powerful object-oriented programming approach to software development it puts at our disposal. In this chapter, we will put the knowledge we gained from the previous chapter to good use by developing an Android application.

In this chapter, you will learn about the following topics:

- Android application components
- Views
- View groups
- Layout constraints
- Implementing layouts with XML
- String and dimension resources
- Handling input events

We will learn these topics through a hands-on approach by implementing the layouts and components of a classic game, Tetris, in the form of an Android application. As we are developing the game in the form of an Android application, before proceeding further, it is imperative that we do a brief overview of the Android operating system.

Android – an overview

Android is a Linux-based mobile operating system developed and maintained by Google and created primarily to power smart mobile devices such as mobile phones and tablets. The primary interface to interact with the Android operating system is **Graphic User Interface (GUI)**-based. Users of devices powered by Android manipulate and interact with the operating system environment primarily via a visual touch-based interface by performing gestures such as taps and swipes on the display.

Software can be installed on the Android OS in the form of apps. An app is an application that runs within an environment and performs one or more tasks for the achievement of a goal or a collection of goals. The ability to install applications on mobile devices presented a huge opportunity to both users and application developers. Users take advantage of the features provided by apps to achieve day-to-day goals and developers take advantage of the demand for software applications by developing apps that meet user needs and perhaps make a profit.

To developers, Android provides a vast array of tools and utilities for the development of high-performance applications. These applications can target different markets such as recreation, enterprise, and e-commerce. Applications can also come in the form of games.

Over the course of this chapter, we will explore a number of these tools and utilities, which are provided by the Android application framework, in more detail.

Application components

The Android application framework provides us with a number of components we can utilize to build a user interface for the *Tetris* application. A component in Android is a reusable program template or object that can be used to define aspects of an application. Some important components provided by the Android application framework are:

- Activities
- Intents
- Intent filters
- Fragments
- Services
- Loaders
- Content providers

Activities

An activity is an Android component that is central to the implementation of application flow and component-to-component interaction. Activities are implemented in the form of classes. An instance of an activity is used by the Android system for code initiation.

An activity is important in the creation of applications' user interfaces. It provides a window upon which user interface elements can be drawn. Simply put, application screens are created with the use of activities in mind.

Intents

An intent facilitates inter-activity communication. Intents can be considered messengers within an Android application. They are messaging objects that are used to request actions from application components. Intents can be used for actions, such as requesting the start of an activity and delivering broadcasts, within the Android system environment.

There are two types of intents. These are:

- Implicit intents
- Explicit intents

Implicit intents: These are messenger objects that do not specifically identify an application component to perform an action, but specify an action to be performed and allow a component that may exist in another application to perform the action. The components that can handle an action requested implicitly are identified by the Android system.

Explicit intents: These intents specify explicitly the application component that should perform an action. These can be used to perform actions, such as starting an activity, within your application:

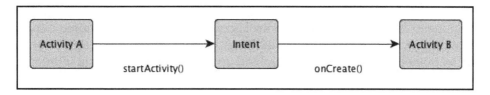

Intent filters

An intent filter is a declaration in the application manifest file that specifies the type of intent that a component would like to receive. This is useful in a number of cases, such as a scenario in which you want an activity in your application to handle a specific action requested by components in another application. For this case, an intent filter can be declared in the application manifest for the activity you want to handle the external request. If you do not want an activity to handle implicit intents, you simply do not declare an intent filter for it.

Fragments

A fragment is an application component that represents a part of a user interface that exists within an activity. Similar to an activity, a fragment possesses a layout that can be modified and is drawn on the activity window.

Services

Unlike most other components, a service does not provide a user interface. Services are used to perform background processes in an application. A service does not need the application that created it to be in the foreground to run.

Loaders

A loader is a component that enables the loading of data from a data source, such as a content provider, for later display in an activity or fragment.

Content providers

These components help an application control access to data resources stored either within the application or within another app. In addition, a content provider facilitates the sharing of data with another application via an exposed application programming interface:

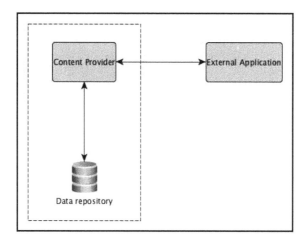

Understanding Tetris

Before attempting to develop the *Tetris* game as an Android application, we need to understand the game, its rules, and its constraints.

Tetris is a matching puzzle video game that makes use of tiles. The name *Tetris* is derived from the words tetra – the Greek numerical prefix for four – and tennis. The tiles in Tetris combine to make up tetrominoes which are geometric shapes composed of four squares connected orthogonally:

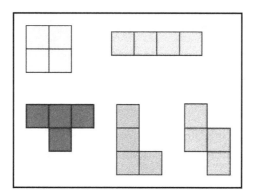

Tetris tetrominoes

In Tetris, a random sequence of tetrominoes fall down upon a playing field. These tetrominoes can be manipulated by the player. A number of motions can be performed on each tetromino piece. Pieces can be moved to the left, to the right, and rotated. In addition, the speed of descent of each piece can be sped up. The objective of the game is to create an uninterrupted horizontal line of ten cells with the descending pieces. When such a line is created, the line is cleared.

Now that we have an understanding of how Tetris works, let's get into the specifics in order to build the user interface of the application.

Creating the user interface

As previously mentioned, the user interface is the means by which the user of an application interacts with the app. It cannot be overstated how important the user interface of an application is. Before embarking on the process of actually coding a user interface, it may be helpful to make a graphical representation of the UI to be implemented. This can be done with different tools, such as Photoshop, but for this case, a simple sketch is sufficient:

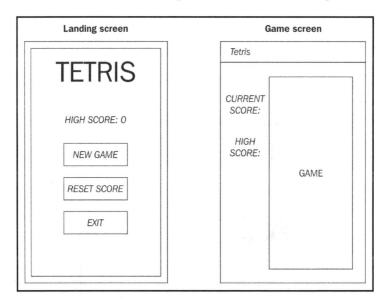

From the preceding sketch, we can see that we will need two distinct screens in this application: the landing screen and the game screen where the actual gameplay will happen. These two screens will require two separate activities. We'll call these two activities MainActivity and GameActivity.

MainActivity will serve as the entry point of our application. It will contain the user interface and all logic pertaining to our landing screen. As we can observe in our sketch, the UI of the landing screen contains the application title, a view that shows the user their current high score, and three buttons performing different actions. The **NEW GAME** button, as the name implies, will lead the user to the activity in which the gaming takes place. **RESET SCORE** will reset the score of the user to zero and **EXIT** will close the application.

GameActivity will be the programmatic template of the game screen. In this activity, we will create the views and logical interactions between the user and the game. The UI of this activity contains an action bar with the title of the application displayed on it, two text views that display the current score of the user and their high score, and a layout element in which the Tetris gameplay will happen.

Implementing the layouts

Now that we know the activities that are needed in this application and have a rough idea of how we want our user interface to look when viewed by the user, we can get into the actual implementation of the user interface.

Create a new Android project in Android Studio and give it the name Tetris no activity. Once the IDE window opens, you will notice that the project structure is similar to the one in Chapter 1, *The Fundamentals*.

The first thing we need to do is add `MainActivity` to our project. We want `MainActivity` to be an empty activity. We can add `MainActivity` to our project by right-clicking on our source package and selecting **New** | **Activity** | **Empty Activity**:

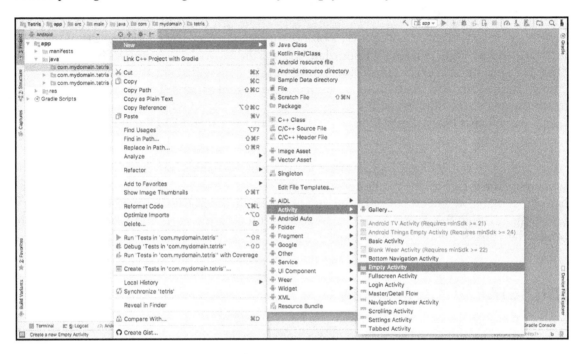

Name the activity `MainActivity`, and ensure the **Generate Layout File**, **Launcher Activity**, and **Backwards Compatibility (AppCompat)** checkboxes are ticked.

Upon the addition of the activity to your project, navigate to its layout resource file. It should look similar to the following code:

```xml
<?xml version="1.0"
encoding="utf-8"?><android.support.constraint.ConstraintLayout
xmlns:android="http://schemas.android.com/apk/res/android"
    xmlns:app="http://schemas.android.com/apk/res-auto"
    xmlns:tools="http://schemas.android.com/tools"
    android:layout_width="match_parent"
    android:layout_height="match_parent"
    tools:context="com.mydomain.tetris.MainActivity">
</android.support.constraint.ConstraintLayout>
```

Line one of the resource file specifies the XML version utilized in the file as well as the character encoding used. utf-8 is used for character encoding in this file. **UTF** stands for **Unicode Transformation Format**. It is an encoding format that can be as compact as **American Standard Code for Information Interchange (ASCII)**—the most common character format for text files—and can contain any Unicode character. The next eight lines define a ConstraintLayout to be rendered in the UI of MainActivity.

Let's consider the ConstraintLayout in a bit more detail before moving forward.

ConstraintLayout

A ConstraintLayout is a type of view group that allows the flexible positioning and resizing of application widgets. Various types of constraints can be used on a ConstraintLayout. Some examples are.

Margins

A margin is a space that exists between two layout elements. When a side margin is set on an element, it is applied to its corresponding layout constraints, if one is available, by adding the margin as a space between the target side and the source side (the side of the element adding the margin):

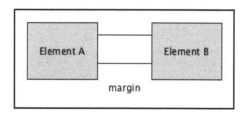

Chains

Chains are constraints that provide group-like behavior in a single axis. The axis may be either horizontal or vertical:

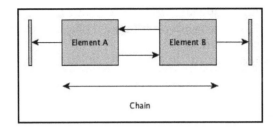

A collection of elements is a chain if they are all connected bidirectionally.

Dimension constraints

These constraints concern the sizes of widgets placed in a layout. Dimension constraints can be set on widgets, and using a `ConstraintLayout`:

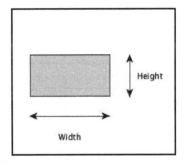

A dimension constraint

The dimensions of a widget can be specified by the use of `android:layout_width` and `android:layout_height`:

```
<TextView
  android:layout_height="16dp"
  android:layout_width="32dp"/>
```

In a number of cases, you may want a widget to have the same dimension as its parent view group. This can be done by assigning the match_parent value to the dimension attribute:

```
<LinearLayout
  android:layout_width="120dp"
  android:layout_height="100dp">
  <TextView
    android:layout_width="match_parent"
    android:layout_height="match_parent"/>
</LinearLayout>
```

Alternatively, if you want a widget's dimensions not to be fixed but rather to wrap the elements contained within it, the wrap_content value should be assigned to the dimension attribute:

```
<TextView
  android:layout_width="wrap_content"
  android:layout_height="wrap_content"
  android:text="I wrap around the content within me"
  android:textSize="15sp"/>
```

Now that we have a better understanding of the ConstraintLayout, as well as widget constraints, let's take another look at our activity_main.xml file:

```
<android.support.constraint.ConstraintLayout
xmlns:android="http://schemas.android.com/apk/res/android"
    xmlns:app="http://schemas.android.com/apk/res-auto"
    xmlns:tools="http://schemas.android.com/tools"
    android:layout_width="match_parent"
    android:layout_height="match_parent"
    tools:context="com.mydomain.tetris.MainActivity">
</android.support.constraint.ConstraintLayout>
```

Looking at the ConstraintLayout element, we can immediately notice that its width and height dimensions have been set to match_parent. This means that the ConstraintLayout dimensions are set to match those of the current window. The attributes that have the xmlns: prefix are used to define XML namespaces. Values set for all XML namespace attributes are namespace URIs. **URI** is short for **Uniform Resource Identifier** and, as the name suggests, it identifies a resource required by the namespace.

The tools:context attribute is typically set to the root element in an XML layout file and specifies the activity that the layout is associated with—in this case, MainActivity.

Now that we understand what's going on in the `activity_main.xml` layout, let's add some layout elements to it. From our sketch, we can see that all layout elements are placed in a vertical arrangement. We can do this with the use of a `LinearLayout`:

```
<android.support.constraint.ConstraintLayout
xmlns:android="http://schemas.android.com/apk/res/android"
    xmlns:app="http://schemas.android.com/apk/res-auto"
    xmlns:tools="http://schemas.android.com/tools"
    android:layout_width="match_parent"
    android:layout_height="match_parent"
    tools:context="com.mydomain.tetris.MainActivity">
    <LinearLayout
        android:layout_width="match_parent"
        android:layout_height="match_parent"
        app:layout_constraintBottom_toBottomOf="parent"
        app:layout_constraintLeft_toLeftOf="parent"
        app:layout_constraintRight_toRightOf="parent"
        app:layout_constraintTop_toTopOf="parent"
        android:layout_marginVertical="16dp"
        android:orientation="vertical">
    </LinearLayout>
</android.support.constraint.ConstraintLayout>
```

As we want the `LinearLayout` to have the same dimensions as its parent, we set both `android:layout_width` and `android:layout_height` to `match_parent`. Next, we specify the edge constraints of the `LinearLayout` using the `app:layout_constraintBottom_toBottomOf`, `app:layout_constraintLeft_toLeftOf`, `app:layout_constraintRight_toRightOff`, and `app:layout_constraintTop_toTopOf` attributes.

- `app:layout_constraintBottom_toBottomOf`: Aligns the bottom edge of the element to the bottom of another
- `app:layout_constraintLeft_toLeftOf`: Aligns the left edge of an element to the left of another
- `app:layout_constraintRight_toRightOf`: Aligns the right edge of an element to the right of another
- `app:layout_constraintTop_toTopOf`: Aligns the top of an element to the top of another.

In this case, all edges of the `LinearLayout` are aligned to the edge of its parent—the `ConstraintLayout`. `android:layout_marginVertical` adds a margin of `16dp` to the top and bottom of the element.

Defining dimension resources

Typically in a layout file, we can have numerous elements that specify the same constraint values to attributes. Such values should be added to a dimensions resource file. Let's go ahead and create a dimensions resource file now. In your application project view, navigate to **res** | **values** and create a new value resource file in the directory with the name `dimens`:

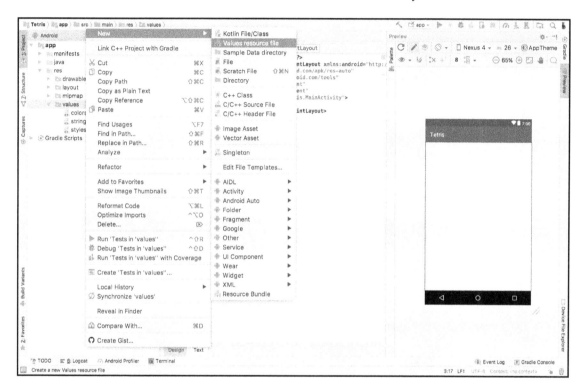

Leave all other file attributes at their default values:

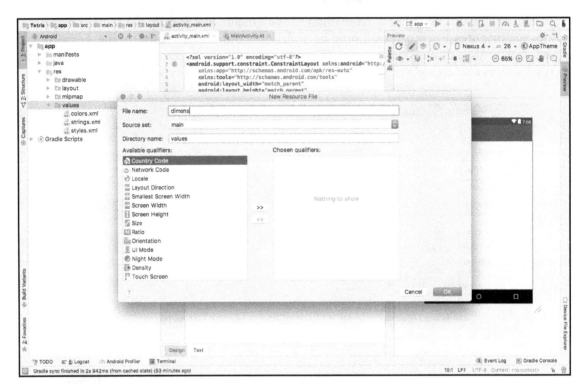

Open the file upon its creation. Its content should be similar to the following code:

```
<?xml version="1.0" encoding="utf-8"?>
<resources></resources>
```

The first line of `dimens.xml` declares the XML version and character encoding used within the file. The second line contains a `<resources>` resources tag. Our dimensions will be declared within this tag. Add a few dimension values, as demonstrated in the following code:

```
<?xml version="1.0" encoding="utf-8"?>
<resources>
    <dimen name="layout_margin_top">16dp</dimen>
    <dimen name="layout_margin_bottom">16dp</dimen>
    <dimen name="layout_margin_start">16dp</dimen>
    <dimen name="layout_margin_end">16dp</dimen>
    <dimen name="layout_margin_vertical">16dp</dimen>
</resources>
```

New dimensions are declared using the `<dimen>` tag. Dimension names should typically be written in snake case. The value for a dimension is added within the opening `<dimen>` and closing `</dimens>` tags.

Now that we have added a few dimensions, we can use them in our linear layout:

```
<android.support.constraint.ConstraintLayout
xmlns:android="http://schemas.android.com/apk/res/android"
    xmlns:app="http://schemas.android.com/apk/res-auto"
    xmlns:tools="http://schemas.android.com/tools"
    android:layout_width="match_parent"
    android:layout_height="match_parent"
    tools:context="com.mydomain.tetris.MainActivity">
    <LinearLayout
        android:layout_width="match_parent"
        android:layout_height="match_parent"
        app:layout_constraintBottom_toBottomOf="parent"
        app:layout_constraintLeft_toLeftOf="parent"
        app:layout_constraintRight_toRightOf="parent"
        app:layout_constraintTop_toTopOf="parent"
        android:layout_marginTop="@dimen/layout_margin_top"
<!-- layout_margin_top dimension reference -->
        android:layout_marginBottom="@dimen/layout_margin_bottom"
<!-- layout_margin_top dimension reference -->
        android:orientation="vertical"
        android:gravity="center_horizontal">
    </LinearLayout>
</android.support.constraint.ConstraintLayout>
```

We've gotten our `LinearLayout` view group set up and now we need to add the required layout views to it. Before we do that, we need to understand the concepts of views and view groups.

Views

A view is a layout element that occupies a set area of the screen and is responsible for drawing and event handling. View is the base class for UI elements or widgets such as text fields, input fields, and buttons. All views extend the View class.

Views can be created within an XML layout in a source file. Consider the following code:

```
<TextView
  android:layout_width="wrap_content"
  android:layout_height="wrap_content"
  android:text="Roll the dice!"/>
```

Besides creating views directly in a layout file, they can also be created programmatically within program files. For example, a text view can be made by creating an instance of the TextView class and passing a context to its constructor. This is demonstrated in the following code snippet:

```
class MainActivity : AppCompatActivity() {
  override fun onCreate(savedInstanceState: Bundle?) {
    super.onCreate(savedInstanceState)
    setContentView(R.layout.activity_main)
    val textView: TextView = TextView(this)
  }
}
```

View groups

A view group is a special kind of view that is capable of containing views. A view group that contains one or more views is commonly referred to as a parent view and the views contained as its children views. A view group is the parent class of several other view containers. Some examples of view groups are LinearLayout, CoordinatorLayout, ConstriantLayout, RelativeLayout, AbsoluteLayout, GridLayout, and FrameLayout.

View groups can be created within an XML layout in a source file:

```
<LinearLayout
  android:layout_width="wrap_content"
  android:layout_height="wrap_content"
  android:layout_marginTop="16dp"
  android:layout_marginBottom="16dp"/>
```

Similar to views, view groups can be created programmatically within component classes. In the following code snippet, a linear layout is made by creating an instance of the LinearLayout class and passing the context of MainActivity to its constructor:

```
class MainActivity : AppCompatActivity() {
  override fun onCreate(savedInstanceState: Bundle?) {
    super.onCreate(savedInstanceState)
    setContentView(R.layout.activity_main)
```

```
    val linearLayout: LinearLayout = LinearLayout(this)
  }
}
```

Having understood the concepts of views and view groups, we can add a few more views to our layout. Text views are added to a layout with the `<TextView>` element and buttons are added with the `<Button>` element:

```
<?xml version="1.0" encoding="utf-8"?>
<android.support.constraint.ConstraintLayout
xmlns:android="http://schemas.android.com/apk/res/android"
    xmlns:app="http://schemas.android.com/apk/res-auto"
    xmlns:tools="http://schemas.android.com/tools"
    android:layout_width="match_parent"
    android:layout_height="match_parent"
    tools:context="com.mydomain.tetris.MainActivity">
  <LinearLayout
        android:layout_width="match_parent"
        android:layout_height="match_parent"
        app:layout_constraintBottom_toBottomOf="parent"
        app:layout_constraintLeft_toLeftOf="parent"
        app:layout_constraintRight_toRightOf="parent"
        app:layout_constraintTop_toTopOf="parent"
        android:layout_marginTop="@dimen/layout_margin_top"
        android:layout_marginBottom="@dimen/layout_margin_bottom"
        android:orientation="vertical">
    <TextView
          android:layout_width="wrap_content"
          android:layout_height="wrap_content"
          android:text="TETRIS"
          android:textSize="80sp"/>
    <TextView
          android:id="@+id/tv_high_score"
          android:layout_width="wrap_content"
          android:layout_height="wrap_content"
          android:text="High score: 0"
          android:textSize="20sp"
          android:layout_marginTop="@dimen/layout_margin_top"/>
    <LinearLayout
          android:layout_width="match_parent"
          android:layout_height="0dp"
          android:layout_weight="1"
          android:orientation="vertical">
      <Button
          android:id="@+id/btn_new_game"
          android:layout_width="wrap_content"
          android:layout_height="wrap_content"
```

```
                android:text="New game"/>
        <Button
                android:id="@+id/btn_reset_score"
                android:layout_width="wrap_content"
                android:layout_height="wrap_content"
                android:text="Reset score"/>
        <Button
                android:id="@+id/btn_exit"
                android:layout_width="wrap_content"
                android:layout_height="wrap_content"
                android:text="exit"/>
    </LinearLayout>
   </LinearLayout>
  </android.support.constraint.ConstraintLayout>
```

As our sketch outlines, we have added two text views to hold the application title and the high score, as well as three buttons to execute the required actions. We have made use of two new attributes. These attributes are `android:id` and `android:layout_weight`. The `android:id` attribute is used to set a unique identifier for an element in a layout. No two elements in the same layout can have the same ID. The `android:layout_weight` attribute is used to specify a precedence value for how much space a view should take in its parent container:

```
<LinearLayout
    android:layout_width="match_parent"
    android:layout_height="match_parent"
    android:orientation="vertical">
  <Button
        android:layout_width="70dp"
        android:layout_height="40dp"
        android:text="Click me"/>
  <View
        android:layout_width="70dp"
        android:layout_height="0dp"
        android:layout_weight="1"/>
</LinearLayout>
```

In the preceding code snippet, two child views are contained by a linear layout. The button explicitly sets both its dimensional constraints to `70dp` and `40dp`. The view, on the other hand, has its width explicitly set to `70dp` and has its height set to `0dp`. As a result of the presence of an `android:layout_weight` attribute set to `1`, the view's height is set to cover all remaining space in the parent view.

Now that we understand fully what is going on in our layout, we can take a look at the layout design preview:

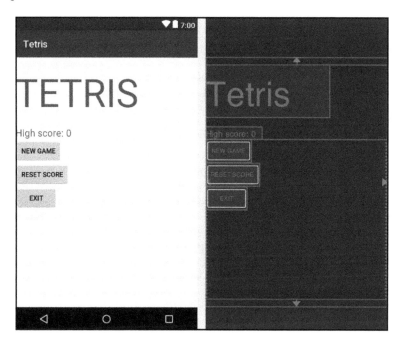

We can see that something seems off. Unlike our sketch, our layout items are not centered but aligned to the right. We can solve this by using the android:gravity attribute in our linear layout view groups. In the following code snippet, we make use of the android:gravity attribute to center layout widgets within both linear layouts:

```xml
<?xml version="1.0" encoding="utf-8"?>
<android.support.constraint.ConstraintLayout
xmlns:android="http://schemas.android.com/apk/res/android"
    xmlns:app="http://schemas.android.com/apk/res-auto"
    xmlns:tools="http://schemas.android.com/tools"
    android:layout_width="match_parent"
    android:layout_height="match_parent"
    tools:context="com.mydomain.tetris.MainActivity">
  <LinearLayout
        android:layout_width="match_parent"
        android:layout_height="match_parent"
        app:layout_constraintBottom_toBottomOf="parent"
        app:layout_constraintLeft_toLeftOf="parent"
        app:layout_constraintRight_toRightOf="parent"
        app:layout_constraintTop_toTopOf="parent"
```

```
            android:layout_marginTop="@dimen/layout_margin_top"
            android:layout_marginBottom="@dimen/layout_margin_bottom"
            android:orientation="vertical"
            android:gravity="center">
        <!-- Aligns child elements to the centre of view group   -->
        <TextView
            android:layout_width="wrap_content"
            android:layout_height="wrap_content"
            android:text="TETRIS"
            android:textSize="80sp"/>
        <TextView
            android:id="@+id/tv_high_score"
            android:layout_width="wrap_content"
            android:layout_height="wrap_content"
            android:text="High score: 0"
            android:textSize="20sp"
            android:layout_marginTop="@dimen/layout_margin_top"/>
        <LinearLayout
            android:layout_width="match_parent"
            android:layout_height="0dp"
            android:layout_weight="1"
            android:orientation="vertical"
            android:gravity="center">
        <!-- Aligns child elements to the centre of view group   -->
        <Button
            android:id="@+id/btn_new_game"
            android:layout_width="wrap_content"
            android:layout_height="wrap_content"
            android:text="New game"/>
        <Button
            android:id="@+id/btn_reset_score"
            android:layout_width="wrap_content"
            android:layout_height="wrap_content"
            android:text="Reset score"/>
        <Button
            android:id="@+id/btn_exit"
            android:layout_width="wrap_content"
            android:layout_height="wrap_content"
            android:text="exit"/>
    </LinearLayout>
  </LinearLayout>
</android.support.constraint.ConstraintLayout>
```

As a result of `android:gravity` being set to `center`, widgets are properly aligned as we would like. The effects of applying the `android:gravity` view groups to our layout view groups can be seen in the following screenshot:

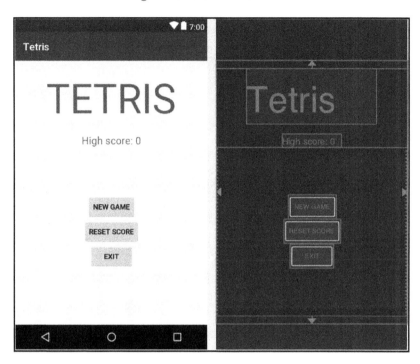

Defining string resources

Up till now, we have been passing hardcoded strings as values to element attributes that require text to be set. This is not best practice and generally should be avoided. Instead, string values should be added in a string resource file.

The default file for string resources is `strings.xml` and this can be found in the `res |
values` directory:

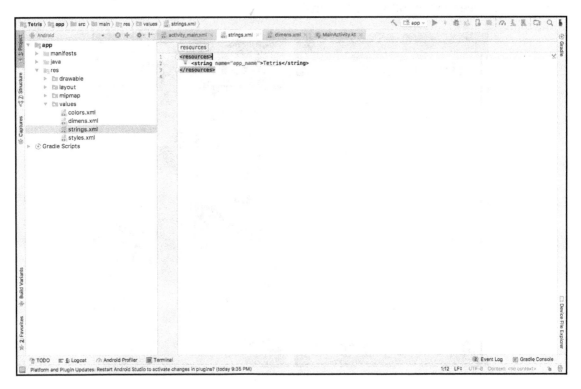

String values are added as string resources using the `<string>` XML tag. We need to add
string resources for all string values we have used thus far. Add the following code to your
string resource file:

```
<resources>
    <string name="app_name">Tetris</string>
    <string name="high_score_default">High score: 0</string>
    <string name="new_game">New game</string>
    <string name="reset_score">Reset score</string>
    <string name="exit">exit</string>
</resources>
```

Now it is necessary to edit our `MainActivity` layout file to exploit these created resources. A string resource can be referenced with `@strings/` prefixing the string resource name. Consider the following code:

```xml
<?xml version="1.0" encoding="utf-8"?>
<android.support.constraint.ConstraintLayout
xmlns:android="http://schemas.android.com/apk/res/android"
    xmlns:app="http://schemas.android.com/apk/res-auto"
    xmlns:tools="http://schemas.android.com/tools"
    android:layout_width="match_parent"
    android:layout_height="match_parent"
    tools:context="com.mydomain.tetris.MainActivity">
  <LinearLayout
        android:layout_width="match_parent"
        android:layout_height="match_parent"
        app:layout_constraintBottom_toBottomOf="parent"
        app:layout_constraintLeft_toLeftOf="parent"
        app:layout_constraintRight_toRightOf="parent"
        app:layout_constraintTop_toTopOf="parent"
        android:layout_marginTop="@dimen/layout_margin_top"
        android:layout_marginBottom="@dimen/layout_margin_bottom"
        android:orientation="vertical"
        android:gravity="center">
    <!-- Aligns child elements to the centre of view group  -->
    <TextView
        android:layout_width="wrap_content"
        android:layout_height="wrap_content"
        android:text="@string/app_name"
        android:textAllCaps="true"
        android:textSize="80sp"/>
    <TextView
        android:id="@+id/tv_high_score"
        android:layout_width="wrap_content"
        android:layout_height="wrap_content"
        android:text="@string/high_score_default"
        android:textSize="20sp"
        android:layout_marginTop="@dimen/layout_margin_top"/>
    <LinearLayout
        android:layout_width="match_parent"
        android:layout_height="0dp"
        android:layout_weight="1"
        android:orientation="vertical"
        android:gravity="center">
      <!-- Aligns child elements to the centre of view group  -->
      <Button
        android:id="@+id/btn_new_game"
        android:layout_width="wrap_content"
```

```
            android:layout_height="wrap_content"
            android:text="@string/new_game"/>
        <Button
            android:id="@+id/btn_reset_score"
            android:layout_width="wrap_content"
            android:layout_height="wrap_content"
            android:text="@string/reset_score"/>
        <Button
            android:id="@+id/btn_exit"
            android:layout_width="wrap_content"
            android:layout_height="wrap_content"
            android:text="@string/exit"/>
    </LinearLayout>
  </LinearLayout>
</android.support.constraint.ConstraintLayout>
```

Handling input events

In the cycle of a user's interaction with an application, a means by which the user can provide some form of input for the execution of a process is by interacting with a widget. These inputs can be captured with events. In Android applications, events are captured from the specific view object that the user interacts with. The required structures and procedures necessary for input event handling are provided by the View class.

Event listeners

An event listener is a procedure in an application program that waits for a UI event to occur. There are many types of events that can be emitted within an application. Some common events are click events, touch events, long click events, and text change events.

In order to capture a widget event and perform an action upon its occurrence, a listener for the event must be set to the view. This can be achieved by invoking a view's set. Listener() method and passing either a lambda or a reference to a function to the method invocation.

The following example demonstrates the capturing of a click event done on a button. A lambda is passed to the setOnClickListener method of the view class:

```
val button: Button = findViewById<Button>(R.id.btn_send)
button.setOnClickListener {
  // actions to perform on click event
}
```

A reference to a function can be passed in place of a lambda:

```
class MainActivity : AppCompatActivity() {
  override fun onCreate(savedInstanceState: Bundle?) {
    super.onCreate(savedInstanceState)
    setContentView(R.layout.activity_main)
    val btnExit: Button = findViewById<Button>(R.id.btn_exit)
    btnExit.setOnClickListener(this::handleExitEvent)
  }
  fun handleExitEvent(view: View) {
    finish()
  }
}
```

Many listener setter methods are available in the view class. Some examples are:

- `setOnClickListener()`: Sets a function to be invoked upon the click of a view
- `setOnContextClickListener()`: Sets a function to be invoked upon the context click of a view
- `setOnCreateContextMenuListener()`: Sets a function to be invoked upon the creation of a view's context menu
- `setOnDragListener()`: Sets a function to be invoked on the occurrence of a drag event on a view
- `setOnFocusChangeListener()`: Sets a function to be called on the focus change of a view
- `setOnHoverChangeListener()`: Sets a function to be called when a hover event occurs on a view
- `setOnLongClickListener()`: Sets a function to be invoked on the occurrence of a long click event on a view
- `setOnScrollChangeListener()`: Sets a function to be invoked when the scroll positions (X or Y) of a view change

 An event listener is a procedure in an application program that waits for a UI event to occur.

As we now have a good understanding of how to handle input events, we can go on to implement some logic in our `MainActivity`.

The main activity screen contains an app bar. We need to hide this layout element as our view does not require it:

Appbar

An app bar is also referred to as an action bar. Action bars are instances of the `ActionBar` class. The instance of the action bar widget in a layout can be retrieved via the `supportActionBar` accessor variable. The following code retrieves the action bar, and hides it if a null reference is not returned:

```
package com.mydomain.tetris
import android.support.v7.app.AppCompatActivity
import android.os.Bundle
import android.support.v7.app.ActionBar
import android.view.View
import android.widget.Button

class MainActivity : AppCompatActivity() {
  override fun onCreate(savedInstanceState: Bundle?) {
    super.onCreate(savedInstanceState)
    setContentView(R.layout.activity_main)
    val appBar: ActionBar? = supportActionBar

    if (appBar != null) {
      appBar.hide()
    }
  }
}
```

Though the preceding code performs what is necessary, its length can be reduced considerably by exploiting Kotlin's type-safe system, which is as follows:

```
package com.mydomain.tetris
import android.support.v7.app.AppCompatActivity
import android.os.Bundle
import android.view.View
import android.widget.Button

class MainActivity : AppCompatActivity() {
```

```
override fun onCreate(savedInstanceState: Bundle?) {
    super.onCreate(savedInstanceState)
    setContentView(R.layout.activity_main)
    supportActionBar?.hide()
  }
}
```

If `supportActionBar` is not a null object reference, the `hide()` method will be invoked if nothing else happens. This will prevent the raising of a null pointer exception.

We need to create object references for the widgets that exist in our layouts. This is necessary for many reasons, such as listener registration. Object references of a view can be retrieved by passing the resource ID of the view to `findViewById()`. We add object references to `MainActivity` (existing in the `MainActivity.kt` file) in the following code snippet:

```
package com.mydomain.tetris
import android.support.v7.app.AppCompatActivity
import android.os.Bundle
import android.view.View
import android.widget.Button
import android.widget.TextView

class MainActivity : AppCompatActivity() {

  var tvHighScore: TextView? = null

  override fun onCreate(savedInstanceState: Bundle?) {
    super.onCreate(savedInstanceState)
    setContentView(R.layout.activity_main)
    supportActionBar?.hide()

    val btnNewGame = findViewById<Button>(R.id.btn_new_game)
    val btnResetScore = findViewById<Button>(R.id.btn_reset_score)
    val btnExit = findViewById<Button>(R.id.btn_exit)
    tvHighScore = findViewById<TextView>(R.id.tv_high_score)
  }
}
```

Now that we have object references to our user interface elements in place, we need to handle some of their events. We must set click listeners for all buttons in the layout (there's no point having a button that does nothing when clicked, after all).

As we stated earlier on, the **New Game** button has the sole task of navigating the user to the game activity (where game play takes place). In order to do this, we will need to utilize an explicit intent. Add a private function containing the logic to be executed on the click of the **New Game** button to `MainActivity` (in the `MainActivity.kt` file) and set a reference to the function via `setOnClickListener()` invocation:

```kotlin
package com.mydomain.tetris
import android.support.v7.app.AppCompatActivity
import android.os.Bundle
import android.view.View
import android.widget.Button
import android.widget.TextView

class MainActivity : AppCompatActivity() {

  var tvHighScore: TextView? = null

  override fun onCreate(savedInstanceState: Bundle?) {
    super.onCreate(savedInstanceState)
    setContentView(R.layout.activity_main)
    supportActionBar?.hide()
    val btnNewGame = findViewById<Button>(R.id.btn_new_game)
    val btnResetScore = findViewById<Button>(R.id.btn_reset_score)
    val btnExit = findViewById<Button>(R.id.btn_exit)
    tvHighScore = findViewById<TextView>(R.id.tv_high_score)

    btnNewGame.setOnClickListener(this::onBtnNewGameClick)
  }

  private fun onBtnNewGameClick(view: View) {     }
}
```

Create a new empty activity and name it `GameActivity`. Once the activity is created, we can utilize an intent to launch the activity on the click of the **New Game** button, as shown in the following code:

```kotlin
private fun onBtnNewGameClick(view: View) {
  val intent = Intent(this, GameActivity::class.java)
  startActivity(intent)
}
```

The first line of the function body creates a new instance of the Intent class and passes the current context and the required activity class to the constructor. Notice we passed `this` as the first argument to the constructor. The `this` keyword is used to refer to the current instance in which `this` is called. Hence, we are actually passing the current activity (`MainActivity`) as the first argument to the constructor. At this point, you might be asking why we are passing an activity as the first argument of the `Intent` constructor when it requires a context as its first argument. This is because all activities are extensions of the `Context` abstract class. Hence, all activities are in their own rights contexts.

The `startActivity()` method is called to launch an activity from which no result is expected. When an intent is passed as its only argument, it starts an activity from which it expects no result. Go ahead and run the application to observe the effect of the button click.

 `Context` is an abstract class in the Android application framework. The implementation of a context is provided by the Android system. `Context` allows access to application-specific resources. `Context` also allows access to calls for application-level operations such as launching activities, sending broadcasts, and receiving intents.

Now let's implement the following functions for the clicks of the **EXIT** and **RESET SCORE** buttons:

```kotlin
package com.mydomain.tetris
import android.content.Intent
import android.support.v7.app.AppCompatActivity
import android.os.Bundle
import android.view.View
import android.widget.Button
import android.widget.TextView

class MainActivity : AppCompatActivity() {

    var tvHighScore: TextView? = null

    override fun onCreate(savedInstanceState: Bundle?) {
        super.onCreate(savedInstanceState)
        setContentView(R.layout.activity_main)
        supportActionBar?.hide()

        val btnNewGame = findViewById<Button>(R.id.btn_new_game)
        val btnResetScore = findViewById<Button>(R.id.btn_reset_score)
        val btnExit = findViewById<Button>(R.id.btn_exit)
        tvHighScore = findViewById<TextView>(R.id.tv_high_score)
        btnNewGame.setOnClickListener(this::onBtnNewGameClick)
        btnResetScore.setOnClickListener(this::onBtnResetScoreClick)
```

```
        btnExit.setOnClickListener(this::onBtnExitClick)
    }

    private fun onBtnNewGameClick(view: View) {
        val intent = Intent(this, GameActivity::class.java)
        startActivity(intent)
    }

    private fun onBtnResetScoreClick(view: View) {}

    private fun onBtnExitClick(view: View) {
        System.exit(0)
    }
}
```

The call to `System.exit()` in the `onBtnExitClick` function stops further execution of the program and exits it when the 0 integer is passed as its argument. The last thing we need to do concerning handling click events is to implement the logic to perform the reset of high scores. To do this, we need to implement some logic for data storage first to store the high score. We will do this using `SharedPreferences`.

Working with SharedPreferences

`SharedPreferences` is an interface for storing, accessing, and modifying data. The `SharedPreferences` APIs enable data storage in sets of key-value pairs.

We will set up a simple utility to handle our data storage needs for this app utilizing the `SharedPreferences` interface. Create a package in the project's source directory with the name `storage` (right-click the source directory and select **New** | **Package**):

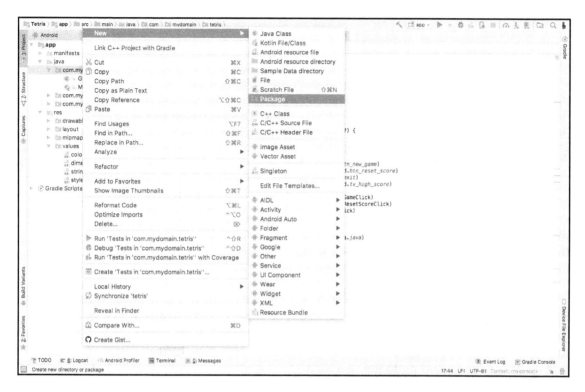

Next, create a new Kotlin class named `AppPreferences` within the `storage` package. Type the following code into the class file:

```
package com.mydomain.tetris.storage
import android.content.Context
import android.content.SharedPreferences

class AppPreferences(ctx: Context) {

  var data: SharedPreferences = ctx.getSharedPreferences
                            ("APP_PREFERENCES", Context.MODE_PRIVATE)

  fun saveHighScore(highScore: Int) {
    data.edit().putInt("HIGH_SCORE", highScore).apply()
  }

  fun getHighScore(): Int {
    return data.getInt("HIGH_SCORE", 0)
  }

  fun clearHighScore() {
```

```
      data.edit().putInt("HIGH_SCORE", 0).apply()
   }
}
```

In the preceding code snippet, a `Context` is required to be passed to the class' constructor upon creation of an instance of the class. Context provides access to the `getSharedPreferences()` method, which retrieves a specified preference file. The preference file is identified by the name in the string passed as the `getSharedPreferences()` method's first argument.

The `saveHighScore()` function takes an integer – the high score to be saved – as its only argument. `data.edit()` returns an `Editor` object that permits the modification of a preference file. The editor's `putInt()` method is called in order to store an integer within the preference file. The first argument passed to `putInt()` is a string representing the key that will be used to access the stored value. The second argument to the method is the integer to be stored – in this case, the high score.

`getHighScore()` returns the high score by calling `data.getInt()`. `getInt()` is a function implemented by `SharedPreferences` that provides read access to a stored integer value. `HIGH_SCORE` is the unique identifier of the value to be retrieved. The 0 passed to the function's second argument specifies a default value to be returned in the scenario that no value corresponding to the specified key exists.

`clearHighScore()` resets the high score to zero by simply overwriting the value corresponding to the `HIGH_SCORE` key with 0.

Now that we have our `AppPreferences` utility class in place, we can finish up the `onBtnResetScoreClick()` function in `MainActivity`:

```
private fun onBtnResetScoreClick(view: View) {
   val preferences = AppPreferences(this)
   preferences.clearHighScore()
}
```

Now when the high score reset button is clicked, the high score is reset to zero. You'll want to give the user some sort of feedback when such actions occur. We can utilize a `Snackbar` to provide this user feedback.

In order to use the `Snackbar` class within an Android application, the Android design support library dependency must be added to the module-level Gradle build script. Do this by adding the following line of code under the dependencies closure of `build.gradle`:

```
implementation 'com.android.support:design:26.1.0'
```

After you have added the line, your module-level `build.gradle` script should look similar to the following:

```
apply plugin: 'com.android.application'
apply plugin: 'kotlin-android'
apply plugin: 'kotlin-android-extensions'

android {
  compileSdkVersion 26
  buildToolsVersion "26.0.1"
  defaultConfig {
    applicationId "com.mydomain.tetris"
    minSdkVersion 15
    targetSdkVersion 26
    versionCode 1
    versionName "1.0"
    testInstrumentationRunner "android.support.test.runner
                              .AndroidJUnitRunner"
  }
  buildTypes {
    release {
      minifyEnabled false
      proguardFiles getDefaultProguardFile('proguard-android.txt'),
                   'proguard-rules.pro'
    }
  }
}

dependencies {
  implementation fileTree(dir: 'libs', include: ['*.jar'])
  implementation "org.jetbrains.kotlin:
                 kotlin-stdlib-jre7:$kotlin_version"
  implementation 'com.android.support:appcompat-v7:26.1.0'
  implementation 'com.android.support.constraint:
                 constraint-layout:1.0.2'
  testImplementation 'junit:junit:4.12'
  androidTestImplementation 'com.android.support.test:runner:1.0.1'
  androidTestImplementation 'com.android.support.test.espresso:espresso-
core:3.0.1'
  implementation 'com.android.support:design:26.1.0'
  // adding android design support library
}
```

After making the alterations, sync your project by clicking **Sync Now** on the flash message that appears within the editor window, as shown in the following screenshot:

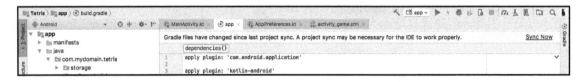

Without further ado, let's modify our `onBtnResetClick()` to provide user feedback in the form of a `Snackbar` after a score reset has been performed by using the following code:

```
private fun onBtnResetScoreClick(view: View) {
    val preferences = AppPreferences(this)
    preferences.clearHighScore()
    Snackbar.make(view, "Score successfully reset",
                Snackbar.LENGTH_SHORT).show()
}
```

Clicking on **RESET SCORE** successfully resets the high score of the player as shown in the following screenshot:

Before moving further, you'll want to update the text displayed in the high score text view of the `MainActivity` layout to reflect the reset score. This can be done by changing the text contained in the text view as follows:

```
private fun onBtnResetScoreClick(view: View) {
  val preferences = AppPreferences(this)
  preferences.clearHighScore()
  Snackbar.make(view, "Score successfully reset",
               Snackbar.LENGTH_SHORT).show()
  tvHighScore?.text = "High score: ${preferences.getHighScore()}"
}
```

Implementing the game activity layout

So far, we have successfully created the layout for main activity. Before we conclude this chapter, it is imperative we create the layout for `GameActivity` as well. Go ahead and open `activity_game.xml` and add the following code to it:

```xml
<?xml version="1.0" encoding="utf-8"?>
<android.support.constraint.ConstraintLayout
xmlns:android="http://schemas.android.com/apk/res/android"
    xmlns:app="http://schemas.android.com/apk/res-auto"
    xmlns:tools="http://schemas.android.com/tools"
    android:layout_width="match_parent"
    android:layout_height="match_parent"
    tools:context="com.mydomain.tetris.GameActivity">
  <LinearLayout
        android:layout_width="match_parent"
        android:layout_height="match_parent"
        android:orientation="horizontal"
        android:weightSum="10"
        android:background="#e8e8e8">
  <LinearLayout
        android:layout_width="wrap_content"
        android:layout_height="match_parent"
        android:orientation="vertical"
        android:gravity="center"
        android:paddingTop="32dp"
        android:paddingBottom="32dp"
        android:layout_weight="1">
    <LinearLayout
        android:layout_width="wrap_content"
        android:layout_height="0dp"
        android:layout_weight="1"
        android:orientation="vertical"
        android:gravity="center">
```

```xml
        <TextView
            android:layout_width="wrap_content"
            android:layout_height="wrap_content"
            android:text="@string/current_score"
            android:textAllCaps="true"
            android:textStyle="bold"
            android:textSize="14sp"/>
        <TextView
            android:id="@+id/tv_current_score"
            android:layout_width="wrap_content"
            android:layout_height="wrap_content"
            android:textSize="18sp"/>
        <TextView
            android:layout_width="wrap_content"
            android:layout_height="wrap_content"
            android:layout_marginTop="@dimen/layout_margin_top"
            android:text="@string/high_score"
            android:textAllCaps="true"
            android:textStyle="bold"
            android:textSize="14sp"/>
        <TextView
            android:id="@+id/tv_high_score"
            android:layout_width="wrap_content"
            android:layout_height="wrap_content"
            android:textSize="18sp"/>
    </LinearLayout>
    <Button
            android:id="@+id/btn_restart"
            android:layout_width="wrap_content"
            android:layout_height="wrap_content"
            android:text="@string/btn_restart"/>
    </LinearLayout>
    <View
            android:layout_width="1dp"
            android:layout_height="match_parent"
            android:background="#000"/>
    <LinearLayout
            android:layout_width="0dp"
            android:layout_height="match_parent"
            android:layout_weight="9">

    </LinearLayout>
  </LinearLayout>
</android.support.constraint.ConstraintLayout>
```

Most view attributes used in this layout have already previously been used and as such do not need further explanation. The only exceptions are the `android:background` and `android:layout_weightSum` attributes.

The `android:background` attribute is used to set the background color of a view or view group. `#e8e8e8` and `#000` were passed as values in the two instances where `android:background` is used in the layout. `#e8e8e8` is the hex color code for gray and `#000` the hex code for black.

`android:layout_weightSum` defines the maximum weight sum in a view group and is calculated as the sum of the `layout_weight` values of all child views in a view group. The first linear layout in `activity_game.xml` declares the weight sum of all child views to be `10`. As such, the immediate children of the linear layout have layout weights of `1` and `9`, respectively.

We made use of three string resources that we have not previously added to our string resources file. Go ahead and add the following string resources to `strings.xml`:

```
<string name="high_score">High score</string>
<string name="current_score">Current score</string>
<string name="btn_restart">Restart</string>
```

Finally, we have to add some simple logic to game activity for the population of the high score and current score text views, as follows:

```
package com.mydomain.tetris

import android.os.Bundle
import android.support.v7.app.AppCompatActivity
import android.widget.Button
import android.widget.TextView
import com.mydomain.tetris.storage.AppPreferences

class GameActivity: AppCompatActivity() {

  var tvHighScore: TextView? = null
  var tvCurrentScore: TextView? = null
  var appPreferences: AppPreferences? = null

  public override fun onCreate(savedInstanceState: Bundle?) {
    super.onCreate(savedInstanceState)
    setContentView(R.layout.activity_game)
    appPreferences = AppPreferences(this)

    val btnRestart = findViewById<Button>(R.id.btn_restart)
```

```
    tvHighScore = findViewById<TextView>(R.id.tv_high_score)
    tvCurrentScore = findViewById<TextView>(R.id.tv_current_score)

    updateHighScore()
    updateCurrentScore()
}

private fun updateHighScore() {
    tvHighScore?.text = "${appPreferences?.getHighScore()}"
}

private fun updateCurrentScore() {
    tvCurrentScore?.text = "0"
}
}
```

In the preceding code snippet, object references to layout view elements are created. In addition, we declare the updateHighScore() and updateCurrentScore() functions. These two functions are invoked on the creation of the view. They set the default scores displayed in the current score and high score text views declared in the layout file.

Save the changes made to the project and build and run the application. Click on the **New Game** button once the application starts to view the layout we just created:

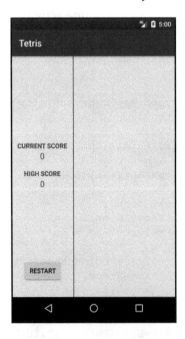

The right-hand side of the layout that contains no content is the area in which the Tetris game play will happen. We will implement this in chapter 3: Implementing Tetris Logic and Functionality. The final thing we must understand before moving to the next chapter is the app manifest.

The app manifest

The app manifest is an XML file that is present in every Android application. It is located in the manifests of an application's root folder. The manifest file holds crucial information pertaining to an application on the Android operating system. The information contained in an application's androidManifest.xml must be read by the Android system before an application can be run. Some of the information that must be registered in the app manifest are:

- The Java package name for the application
- The activities present in the application
- Services that are used in the application
- Intent filters that direct implicit intents to an activity
- Descriptions of the broadcast receivers used in the application
- Data pertaining to content providers present in the application
- The classes that implement the various application components
- The permissions that are required by an application

Structure of the app manifest file

The general structure of the androidManifest.xml file is shown in the following code snippet. The snippet contains all possible elements and declarations that can exist in the manifest file:

```xml
<?xml version="1.0" encoding="utf-8"?>
<manifest>
  <uses-permission />
  <permission />
  <permission-tree />
  <permission-group />
  <instrumentation />
  <uses-sdk />
  <uses-configuration />
  <uses-feature />
```

```
<supports-screens />
<compatible-screens />
<supports-gl-texture />

<application>
  <activity>
    <intent-filter>
      <action />
      <category />
      <data />
    </intent-filter>
    <meta-data />
  </activity>
  <activity-alias>
    <intent-filter>
      . . .
    </intent-filter>
    <meta-data />
  </activity-alias>
  <service>
    <intent-filter>
      . . .
    </intent-filter>
    <meta-data/>
  </service>
  <receiver>
    <intent-filter> . . . </intent-filter>
    <meta-data />
  </receiver>
  <provider>
    <grant-uri-permission />
    <meta-data />
    <path-permission />
  </provider>
  <uses-library />
</application>
</manifest>
```

As can be seen from the preceding code snippet, a vast array of elements can appear in the `manifest` file. Many of these elements will be covered in this book. As a matter of fact, a number of these manifest elements have already been used in our Tetris application. Go ahead and open the `androidManifest.xml` file of Tetris. The contents of the file should be similar to what is contained in the following code snippet:

```
<?xml version="1.0" encoding="utf-8"?>
<manifest xmlns:android="http://schemas.android.com/apk/res/android"
    package="com.mydomain.tetris">
```

```
<application
    android:allowBackup="true"
    android:icon="@mipmap/ic_launcher"
    android:label="@string/app_name"
    android:roundIcon="@mipmap/ic_launcher_round"
    android:supportsRtl="true"
    android:theme="@style/AppTheme">
  <activity android:name=".MainActivity">
    <intent-filter>
      <action android:name="android.intent.action.MAIN" />
      <category android:name="android.intent.category.LAUNCHER" />
    </intent-filter>
  </activity>
  <activity android:name=".GameActivity" />
</application>

</manifest>
```

The elements used in the preceding `manifest` file – in alphabetical order – are as follows:

- `<action>`
- `<activity>`
- `<application>`
- `<category>`
- `<intent-filter>`
- `<manifest>`

<action>

This is used to add an action to an intent filter. The `<action>` element is always a child element to an `<intent-filter>` element. An intent filter should contain one or more of these elements. If no action element is declared for an intent filter, the filter accepts no `Intent` objects. Its syntax is as follows:

```
<action name=""/>
```

The preceding `name` attribute is an attribute that specifies the name of the action being handled.

<activity>

This element declares an activity existing in an application. All activities must be declared in the app manifest in order to be seen by the Android system. <activity> is always placed within a parent <application> element. The following code snippet shows the declaration of an activity within a manifest file using the <activity> element:

```
<activity android:name=".GameActivity" />
```

The name attribute in the preceding code snippet is an attribute that specifies the name of the class that implements the activity being declared.

<application>

This element is the declaration of the application. It contains subelements that declare the components existing in the application. The following code demonstrates the use of <application>:

```
<application
      android:allowBackup="true"
      android:icon="@mipmap/ic_launcher"
      android:label="@string/app_name"
      android:roundIcon="@mipmap/ic_launcher_round"
      android:supportsRtl="true"
      android:theme="@style/AppTheme">
  <activity android:name=".MainActivity">
    <intent-filter>
      <action android:name="android.intent.action.MAIN" />
      <category android:name="android.intent.category.LAUNCHER" />
    </intent-filter>
  </activity>
  <activity android:name=".GameActivity" />
</application>
```

The `<application>` element in the preceding snippet makes use of four attributes. These attributes are:

- `android:allowBackup`: It is used to specify whether the application is allowed to take part in the backup and restore infrastructure. When set to `true`, the application can be backed up by the Android system. Otherwise, if this attribute is set to `false`, no backup of the application will ever be created by the Android system.
- `android:icon`: It specifies the icon resource for the application. It can also be used to specify icon resources for application components.
- `android:label`: It specifies a default label for the application as a whole. It can also be used to specify default labels for application components.
- `android:roundIcon`: It specifies an icon resource to be used when a circular icon resource is required. When an app icon is requested by a launcher, the Android framework returns either `android:icon` or `android:roundIcon`; which is returned depends on the device build configuration. As either can be returned, it is important to specify a resource for both attributes.
- `android:supportsRtl`: It specifies whether an application is willing to support **right-to-left** (**RTL**) layouts. The application is set to support it when this attribute is set to `true`. Otherwise, the application does not support RTL layouts.
- `android:theme`: It specifies a style resource defining a default theme for all activities in the application.

<category>

This element is a child element to `<intent-filter>`. It is used to specify a category name to its parent intent filter component.

<intent-filter>

Specifies the type of intent that activity, service, and broadcast receiver components can respond to. An intent filter is always declared within a parent component with the `<intent-filter>` element.

\<manifest\>

This is the root element of the app manifest file. It contains a single `<application>` element and specifies the `xmlns:android` and `package` attributes.

Summary

In this chapter, we took a closer look at the Android application framework. In the process, we learned about several things, such as the seven fundamental Android app components: activities, intents, intent filters, fragments, services, loaders, and content providers.

In addition, we took a close look at the process of creating a layout, the constraint layout, types of layout constraints that exist, string, dimension resources, views, view groups, and working with `SharedPreferences`. In the next chapter, we will delve further into the world of Tetris and implement the gameplay as well as critical application logic.

Implementing Tetris Logic and Functionality 3

In the previous chapter, we embarked on the development of the classic game *Tetris*. We determined the layout requirements of our application and implemented the layout elements we identified. In the process, we created two activities for the application: `MainActivity` and `GameActivity`. We also implemented the basic characteristics and behaviors of the views, but nothing pertaining to the core gameplay of the app was done. In this chapter, we are going to implement this gameplay. Over the course of this chapter, you will learn about the following topics:

- Exception handling
- The Model-View-Presenter pattern

Implementing the Tetris gameplay

As we are concerned with implementing gameplay, the activity that we will focus on developing further in this chapter is `GameActivity`. The following screenshot shows the final product of all the development done in this chapter:

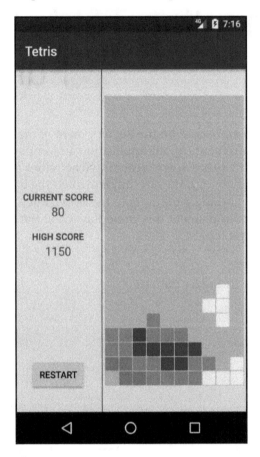

Now that we have an idea of what the final game is going to look like, let's get going with its development.

Under the section *Understanding Tetris* in `Chapter 2`, *Building an Android Application – Tetris* we got to understand that Tetris is a puzzle-matching game that makes use of tiles. These tiles combine to form bigger shapes called tetrominoes. As a reminder, a tetromino is a geometric shape composed of four squares connected orthogonally.

Modeling a tetromino

As tetrominoes are very crucial to the gameplay of Tetris, we must properly model these elements programmatically. In order to do this, let's think of every tetromino piece as a building block. Building blocks have a set of features that they possess. These features can be categorized into characteristics and behaviors.

Characteristics of a block

The following are some characteristics that a block possesses:

- **Shape**: A block has a fixed shape that cannot be changed.
- **Dimensions**: A block possesses dimensional characteristics. These characteristics are height and width.
- **Color**: A block always possesses a color. The color a block possesses is fixed and is maintained throughout the course of its existence.
- **Spatial characteristic**: A block takes up a fixed amount of space.
- **Positional characteristic**: At any given point in time, a block has a position that exists in along two axes, – X and Y.

Behaviors of a block

The main behavior of a block is its ability to experience distinct motions. These motions are translational motion and rotational motion. Translational motion is a type of motion in which a body shifts from one point in space to another. In Tetris, a block can experience leftward, rightward, and downward translational motions. Rotational motion is a type of motion that exists in rigid bodies and follows a curved path. In other words, rotational motion involves the rotation of an object in free space. All blocks in Tetris can be rotated.

Now that we understand the basic characteristics and behaviors of a block, you may be wondering how we can translate them to be relevant to tetrominoes. The truth is no translation of these characteristic features is necessary. All characteristics of a block apply to a tetromino. The only two things to keep in mind are:

- Tetrominoes are made up of four tiles
- All tiles in a tetromino are orthogonally arranged

Having said that, let's get to translating these characteristics into programmatic models. We will start with modeling shape.

Modeling block shape

The approach used to modeling shape varies depending on numerous variables, such as the kind of shape that must be measured and in what spacial dimension the shape is to be modeled. Modeling three-dimensional shapes—all things being equal—is more difficult than modeling two-dimensional shapes. Lucky for us, tetrominoes are two-dimensional in nature. Before we start modeling our shapes programmatically, it is important we know the exact shapes we are attempting to model. There are seven fundamental tetromino pieces that exist in Tetris. These pieces are the O, I, T, L, J, S and Z tetrominos. The following image shows the fundamental tetromino shapes that exist in Tetris:

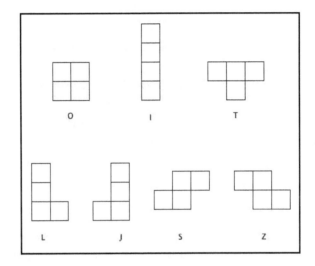

All preceding shapes take up space within the confines of their edges. The area of space covered by a shape can be seen as an outline or a frame. This is similar to how a picture is held within a frame. We need to model this frame that will contain individual shapes. As the shapes being held within the frame are two-dimensional in nature, we will utilize a two-dimensional byte array to hold frame-specific information. A byte is a digital unit of information that generally consists of eight bits. A bit is a binary digit. It is the smallest unit of data in a computer and has a value of either 1 or 0.

The idea is to model the frame of a shape with a two-dimensional array by representing areas covered by the frame with a byte value of 1 and those not covered by it with a value of 0. Take the following frame, for example:

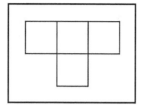

Instead of visualizing it as a whole shape, we can visualize it as a two-dimensional array of bytes possessing two rows and three columns:

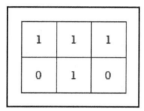

A byte value of 1 is assigned to cells in the array that make up the frame's shape. On the other hand, a byte value of 0 is assigned to cells that are not part of the frame's shape. Modeling this with a class is fairly easy. Firstly, we will need a function that generates the required byte array structure we will use for storing frame bytes. Create a new package within your source package and give it the name `helpers`. Within this package, create a `HelperFunctions.kt` file. This file will contain all helper functions used in the course of the development of this app. Open `HelperFunctions.kt` and type the following code into the file:

```
package com.mydomain.tetris.helpers

fun array2dOfByte(sizeOuter: Int, sizeInner: Int): Array<ByteArray>
        = Array(sizeOuter) { ByteArray(sizeInner) }
```

The preceding code defines a `array2dOfByte()` function, which takes two arguments. The first argument is the desired row number of the array to be generated and the second is the desired column number of the generated byte array. The `array2dOFByte()` method generates and returns a new array with the specified properties. Now that we have our byte array generating helper function set up, let's go ahead and create the `Frame` class. Create a new package within your source package and give it the name `models`. All object models will be packaged within this created package. Within the `models` package, create a `Frame` class in the `Frame.kt` file and type the following code into the file:

```kotlin
package com.mydomain.tetris.models

import com.mydomain.tetris.helpers.array2dOfByte

class Frame(private val width: Int) {
    val data: ArrayList<ByteArray> = ArrayList()

    fun addRow(byteStr: String): Frame {
        val row = ByteArray(byteStr.length)

        for (index in byteStr.indices) {
            row[index] = "${byteStr[index]}".toByte()
        }
        data.add(row)
        return this
    }

    fun as2dByteArray(): Array<ByteArray> {
        val bytes = array2dOfByte(data.size, width)
        return data.toArray(bytes)
    }
}
```

The `Frame` class has two properties: `width` and `data`. Width is an integer property that holds the desired width of the frame to be generated (the number of columns in the frame's byte array). The data property holds an array list of elements in the `ByteArray` value space. We declare two distinct functions, `addRow()` and `get()`. `addRow()` takes a string, converts each individual character of the string into a byte representation, and adds the byte representation into a byte array, after which it adds the byte array to the data list. `get()` converts the data array list into a byte array and returns the array.

Having modeled a suitable frame to hold our block, we can go ahead and model the distinct shapes of possible tetrominoes in the game. In order to do this, we will make use of an enum class. Create a Shape.kt file in the models package before proceeding. We will start by modeling the following simple tetromino shape:

Applying the concept of envisioning frames as a two-dimensional array of bytes, we can envision the frame of the preceding shape as a two-dimensional array of bytes with four rows and a single column with each cell filled with the byte value of 1. With this in mind, let's model the shape. In Shape.kt, create a Shape enum class, as shown in the following code:

```
enum class Shape(val frameCount: Int, val startPosition: Int) {
    Tetromino(2, 2) {
        override fun getFrame(frameNumber: Int): Frame {
            return when (frameNumber) {
                0 -> Frame(4).addRow("1111")
                1 -> Frame(1)
                            .addRow("1")
                            .addRow("1")
                            .addRow("1")
                            .addRow("1")
                else -> throw IllegalArgumentException("$frameNumber is an invalid
                                                        frame number.")
            }
        }
    };
    abstract fun getFrame(frameNumber: Int): Frame
}
```

An enum class is declared by placing the enum keyword before the class keyword. The primary constructor of the preceding Shape enum class takes two arguments. The first argument is frameCount, which is an integer variable that specifies the number of possible frames a shape can be in. The second argument is startPosition, which specifies the intended start position of the shape along the X axis within the gameplay field. Further down the enum class file, a getFrame() function is declared. There's a notable difference between this function and the functions we have declared until now. getFrame() has been declared with the abstract keyword. An abstract function possesses no implementation (thus no body) and is used to abstract a behavior that must be implemented by an extending class. Let's scrutinize the following lines of code within the enum class:

```
Tetromino(2, 2) {
    override fun getFrame(frameNumber: Int): Frame {
        return when (frameNumber) {
            0 -> Frame(4).addRow("1111")
            1 -> Frame(1)
                        .addRow("1")
                        .addRow("1")
                        .addRow("1")
                        .addRow("1")
            else -> throw IllegalArgumentException("$frameNumber is an invalid
                                                    frame number.")
        }
    }
};
```

In the preceding code block, an instance of the enum that provides an implementation of the declared abstract function is being created. The instance's identifier is Tetromino. We passed the integer value 2 as the argument for both the frameCount and startPosition properties of the Tetromino's constructor. In addition, Tetromino provides an implementation for the getFrame() function in its corresponding block by overriding the getFrame() function declared in Shape. Functions are overriden with the override keyword. The implementation of getFrame() in Tetromino takes a frameNumber integer. This frame number determines the frame of Tetromino that will be returned. You may be asking at this point why Tetromino possesses more than one frame. This is simply a result of the possibility of rotation of a tetromino. The single-column tetromino we previously looked at can be rotated either leftwards or rightwards to take the form shown in the following diagram:

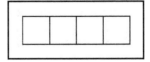

When `frameNumber` passed to `getFrame()` is 0, `getFrame()` returns a `Frame` object that models the frame for the `Tetromino` in its horizontal state, as shown earlier. When `frameNumber` is 1, it returns a frame object modeling the shape in its vertical state.

In the case that `frameNumber` is neither 0 nor 1, an `IllegalArgumentException` is thrown by the function.

 It is important to note that along with being an object, `Tetromino` is a constant. Generally, `enum` classes are used to create constants. An `enum` class is a perfect choice for modeling our tetromino shapes because we have a fixed set of shapes that we need to implement.

Having understood how the `Shape` enum class works, we can model the rest of the possible tetromino shapes as shown in the following code block:

```
enum class Shape(val frameCount: Int, val startPosition: Int) {
```

Let's create a tetromino shape with one frame and a start position of 1. The tetromino modeled here is the square or 'O' shaped tetromino.

```
Tetromino1(1, 1) {
    override fun getFrame(frameNumber: Int): Frame {
        return Frame(2)
            .addRow("11")
            .addRow("11")
    }
},
```

Let's create a tetromino shape with two frames and a start position of 1. The tetromino modeled here is the 'Z' shaped tetromino.

```
Tetromino2(2, 1) {
    override fun getFrame(frameNumber: Int): Frame {
        return when (frameNumber) {
            0 -> Frame(3)
                    .addRow("110")
                    .addRow("011")
            1 -> Frame(2)
                    .addRow("01")
                    .addRow("11")
                    .addRow("10")
            else -> throw IllegalArgumentException("$frameNumber is an invalid
                                                    frame number.")
        }
    }
```

```
    },
```

Let's create a tetromino shape with two frames and a start position of 1. The tetromino modeled here is the 'S' shaped tetromino.

```
Tetromino3(2, 1) {
    override fun getFrame(frameNumber: Int): Frame {
        return when (frameNumber) {
            0 -> Frame(3)
                    .addRow("011")
                    .addRow("110")
            1 -> Frame(2)
                    .addRow("10")
                    .addRow("11")
                    .addRow("01")
            else -> throw IllegalArgumentException("$frameNumber is
                                        an invalid frame number.")
        }
    }
},
```

Let's create a tetromino shape with two frames and a start position of 2. The tetromino modeled here is the 'I' shaped tetromino.

```
Tetromino4(2, 2) {
    override fun getFrame(frameNumber: Int): Frame {
        return when (frameNumber) {
            0 -> Frame(4).addRow("1111")
            1 -> Frame(1)
                    .addRow("1")
                    .addRow("1")
                    .addRow("1")
                    .addRow("1")
            else -> throw IllegalArgumentException("$frameNumber is an
                                        invalid frame number.")
        }
    }
},
```

Let's create a tetromino shape with four frames and a start position of 1. The tetromino modeled here is the 'T' shaped tetromino.

```
Tetromino5(4, 1) {
    override fun getFrame(frameNumber: Int): Frame {
        return when (frameNumber) {
            0 -> Frame(3)
                    .addRow("010")
```

```
                    .addRow("111")
        1 -> Frame(2)
                    .addRow("10")
                    .addRow("11")
                    .addRow("10")
        2 -> Frame(3)
                    .addRow("111")
                    .addRow("010")
        3 -> Frame(2)
                    .addRow("01")
                    .addRow("11")
                    .addRow("01")
        else -> throw IllegalArgumentException("$frameNumber is an
                                    invalid frame number.")
      }
    }
  },
```

Let's create a tetromino shape with four frames and a start position of 1. The tetromino modeled here is the 'J' shaped tetromino.

```
Tetromino6(4, 1) {
  override fun getFrame(frameNumber: Int): Frame {
    return when (frameNumber) {
      0 -> Frame(3)
                  .addRow("100")
                  .addRow("111")
        1 -> Frame(2)
                  .addRow("11")
                  .addRow("10")
                  .addRow("10")
        2 -> Frame(3)
                  .addRow("111")
                  .addRow("001")
        3 -> Frame(2)
                  .addRow("01")
                  .addRow("01")
                  .addRow("11")
        else -> throw IllegalArgumentException("$frameNumber is
                                    an invalid frame number.")
      }
    }
  },
```

Let's create a tetromino shape with four frames and a start position of 1. The tetromino modeled here is the 'L' shaped tetromino.

```
Tetromino7(4, 1) {
    override fun getFrame(frameNumber: Int): Frame {
        return when (frameNumber) {
            0 ->  Frame(3)
                        .addRow("001")
                        .addRow("111")
            1 -> Frame(2)
                        .addRow("10")
                        .addRow("10")
                        .addRow("11")
            2 -> Frame(3)
                        .addRow("111")
                        .addRow("100")
            3 -> Frame(2)
                        .addRow("11")
                        .addRow("01")
                        .addRow("01")
            else -> throw IllegalArgumentException("$frameNumber is
                                        an invalid frame number.")
        }
    }
};

    abstract fun getFrame(frameNumber: Int): Frame
}
```

Having modeled both the block frame and shape, the next thing we must model programmatically is the block itself. We will use this as an opportunity to demonstrate Kotlin's seamless interoperability with Java by implementing the model with Java. Create a new Java class in the models directory (**models** | **New** | **Java Class**) with the name Block. We will start the modeling process by adding instance variables that represent the characteristics of a block. Consider the following code:

```java
package com.mydomain.tetris.models;
import android.graphics.Color;
import android.graphics.Point;

public class Block {
    private int shapeIndex;
    private int frameNumber;
    private BlockColor color;
    private Point position;

    public enum BlockColor {
```

```
PINK(Color.rgb(255, 105, 180), (byte) 2),
GREEN(Color.rgb(0, 128, 0), (byte) 3),
ORANGE(Color.rgb(255, 140, 0), (byte) 4),
YELLOW(Color.rgb(255, 255, 0), (byte) 5),
CYAN(Color.rgb(0, 255, 255), (byte) 6);
BlockColor(int rgbValue, byte value) {
   this.rgbValue = rgbValue;
   this.byteValue = value;
}

private final int rgbValue;
private final byte byteValue;
   }
}
```

In the preceding code block, we add four instance variables: `shapeIndex`, `frameNumber`, `color`, and `position`. `shapeIndex` will hold the index of the shape of the block, `frameNumber` will keep track of the number of frames the block's shape has, `color` will hold the color characteristic of the block, and `position` will be used to keep track of the block's current spatial position in the gaming field.

An `enum` template, `BlockColor`, is added within the `Block` class. This `enum` creates a constant set of `BlockColor` instances, with each possessing `rgbValue` and `byteValue` properties. `rgbValue` is an integer that uniquely identifies an RGB color specified with the `Color.rgb()` method. `Color` is a class provided by the Android application framework and `rgb()` is a class method defined within the `Color` class. The five `Colour.rgb()` calls specify the colors pink, green, orange, yellow, and cyan, respectively.

In `Block`, we made use of the `private` and `public` keywords. These were not added for eye candy; they each have a use. These two keywords, along with the `protected` keyword, are called access modifiers.

> Access modifiers are keywords used to specify access restrictions on classes, methods, functions, variables, and structures. Java has three access modifiers: `private`, `public`, and `protected`. In Kotlin, access modifiers are called visibility modifiers. The available visibility modifiers in Kotlin are `public`, `protected`, `private`, and `internal`.

Private access modifier (private)

Methods, variables, constructors, and structures that are declared private can only be accessed within the declaring class. This is with the exception of private top-level functions and properties that are visible to all members of the same file. Private variables within a class can be accessed from outside the class be declaring getter and setter methods that permit access. Defining setter and getter methods in Java is shown in the following code:

```java
public class Person {
  Person(String fullName, int age) {
    this.fullName = fullName;
    this.age = age;
  }

  private String fullName;
  private int age;

  public String getFullName() {
    return fullName;
  }

  public int getAge() {
    return age;
  }
}
```

In Kotlin, setter and getter creation is as follows:

```kotlin
public class Person(private var fullName: String) {
  var name: String
  get() = fullName
  set(value)  {
    fullName = value
  }
}
```

Using the private access modifier is the main means of information hiding within programs. Information hiding is also known as encapsulation.

Public access modifier (public)

Methods, variables, constructors, and structures declared public can be accessed freely from outside the declaring class. A public class existing in a different package from an accessing class must be imported before it can be used. The following class makes use of the public access modifier:

```
public class Person { .. }
```

Protected access modifier (protected)

Variables, methods, functions, and structures declared protected can be accessed only by classes in the same package as the defining class or by subclasses of their defining class that exist in a separate package:

```
public class Person(private var fullName: String) {
  protected name: String
  get() = fullName
  set(value)  {
    fullName = value
  }
}
```

Internal visibility modifier (internal)

The internal visibility modifier is used to declare a member visible within the same module. A module is a collection of Kotlin files compiled together. A module may be a Maven project, a Gradle source set, and IntelliJ IDEA module, or a set of files compiled with an Ant task invocation. Using the internal modifier is similar to using other visibility modifiers:

```
internal class Person { }
```

Having understood access and visibility modifiers, we can continue with the implementation of the `Block` class. The next thing we need to do is create a constructor for the class that initializes the instance variables we have created to their initial states. Constructor definitions in Java are syntactically different from Kotlin constructor definitions:

```
public class Block {
  private int shapeIndex;
  private int frameNumber;
  private BlockColor color;
  private Point position;
```

Let's see the constructor definition:

```
private Block(int shapeIndex, BlockColor blockColor) {
  this.frameNumber = 0;
  this.shapeIndex = shapeIndex;
  this.color = blockColor;
  this.position = new Point(AppModel.FieldConstants
                       .COLUMN_COUNT.getValue() / 2, 0);
}

public enum BlockColor {
  PINK(Color.rgb(255, 105, 180), (byte) 2),
  GREEN(Color.rgb(0, 128, 0), (byte) 3),
  ORANGE(Color.rgb(255, 140, 0), (byte) 4),
  YELLOW(Color.rgb(255, 255, 0), (byte) 5),
  CYAN(Color.rgb(0, 255, 255), (byte) 6);
  BlockColor(int rgbValue, byte value) {
    this.rgbValue = rgbValue;
    this.byteValue = value;
  }

  private final int rgbValue;
  private final byte byteValue;
}
}
```

Notice that the preceding constructor definition has been given private access. We did this because we do not want this constructor to be accessed outside of `Block`. As we still want other classes to have a means of creating a block instance, we have to define a static method that permits this. We will call this method `createBlock`:

```
public class Block {
  private int shapeIndex;
  private int frameNumber;
  private BlockColor color;
  private Point position;
```

Let's see the Constructor definition:

```
private Block(int shapeIndex, BlockColor blockColor) {
  this.frameNumber = 0;
  this.shapeIndex = shapeIndex;
</span>    this.color = blockColor;
  this.position = new Point( FieldConstants.COLUMN_COUNT
                       .getValue()/2, 0);
}
```

```
public static Block createBlock() {
  Random random = new Random();
  int shapeIndex = random.nextInt(Shape.values().length);
  BlockColor blockColor = BlockColor.values()
      [random.nextInt(BlockColor.values().length)];

  Block block = new Block(shapeIndex, blockColor);
  block.position.x = block.position.x - Shape.values()
      [shapeIndex].getStartPosition();
  return block;
}

public enum BlockColor {
  PINK(Color.rgb(255, 105, 180), (byte) 2),
  GREEN(Color.rgb(0, 128, 0), (byte) 3),
  ORANGE(Color.rgb(255, 140, 0), (byte) 4),
  YELLOW(Color.rgb(255, 255, 0), (byte) 5),
  CYAN(Color.rgb(0, 255, 255), (byte) 6);
  BlockColor(int rgbValue, byte value) {
    this.rgbValue = rgbValue;
    this.byteValue = value;
  }

  private final int rgbValue;
  private final byte byteValue;
}
}
```

createBlock() randomly selects the index of a tetromino shape in the Shape enum class and a BlockColor and assigns two randomly selected values to shapeIndex and blockColor. A new Block instance is created with the two values passed as arguments and the position of the block along the *X* axis is set. Lastly, createBlock() returns the created and initialized block.

We need to add a few getter and setter methods to Block. These methods will give access to crucial properties of instances of the block. Add the following methods to the Block class:

```
public static int getColor(byte value) {
  for (BlockColor colour : BlockColor.values()) {
    if (value == colour.byteValue) {
      return colour.rgbValue;
    }
  }
  return -1;
}
```

```java
public final void setState(int frame, Point position) {
  this.frameNumber = frame;
  this.position = position;
}

@NonNull
public final byte[][] getShape(int frameNumber) {
  return Shape.values()[shapeIndex].getFrame(frameNumber).as2dByteArray();
}

public Point getPosition() {
  return this.position;
}

public final int getFrameCount() {
  return Shape.values()[shapeIndex].getFrameCount();
}

public int getFrameNumber() {
  return frameNumber;
}

public int getColor() {
  return color.rgbValue;
}

public byte getStaticValue() {
  return color.byteValue;
}
```

@NonNull is an annotation provided by the Android application framework that denotes that a field, parameter, or method return can never be null. In the preceding code snippet, it is used in the line prior to the getShape() method definition to denote that the method cannot return a null value.

 In Java, an annotation is a form of metadata that can be added to Java source code. Annotations can be used on classes, methods, variables, parameters, and packages. Annotations can also be declared and used in Kotlin.

The @NotNull annotation exists in the android.support.annotation package. Add the package import to the package imports at the top of Block.java:

```java
import android.support.annotation.NonNull;
```

There's one last thing we should take care of in the `Block` class before moving on. In the final line of the `Block` constructor, the position of the current block instance's position instance variable is set as follows:

```
this.position = new Point(FieldConstants.COLUMN_COUNT.getValue()/2, 0);
```

The `10` is the column count of the field in which the tetrominoes will be generated. This is a constant value that will be used several times within the code for this application, and as such, is best declared as a constant. Create a package named constants in the base application source package and add a new Kotlin file with the name `FieldConstants` to the package. Next, add constants for the number of columns and rows that the playing field will possess. The field should possess ten columns and twenty rows:

```
enum class FieldConstants(val value: Int) {
   COLUMN_COUNT(10), ROW_COUNT(20);
}
```

Import the package with the `FieldConstants` enum class into `Block.java` and replace the `10` integer with the constant value of the `COLUMN_COUNT`:

```
this.position = new Point( FieldConstants.COLUMN_COUNT.getValue()/2, 0);
```

That's it! We are done with the programmatic modeling of the `Block` class.

Creating the application model

Until now, we have been concerned with modeling specific components that make up tetromino blocks. Now it is time to concern ourselves with defining application logic. We will create an application model to implement the necessary Tetris gameplay logic, as well as to serve as an intermediary interface between views and the block components we have created.

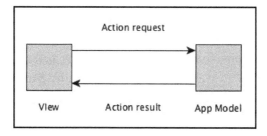

A view will send a request for the performance to the application model, and the model will execute the action if it is valid and send feedback to the view. Similar to the models we have created so far, we need a separate class file for the application model. Go ahead and create a new Kotlin file named `AppModel.kt` and add a class named `AppModel` to the file with imports for `Point`, `FieldConstants`, `array2dOfByte`, and `AppPreferences`:

```
package com.mydomain.tetris.models

import android.graphics.Point
import com.mydomain.tetris.constants.FieldConstants
import com.mydomain.tetris.helpers.array2dOfByte
import com.mydomain.tetris.storage.AppPreferences

class AppModel
```

Some functions of `AppModel` are to keep track of the current score, the `tetris` gameplay field state, the current block, the current state of the game, the current status of the game, and the motions being experienced by the current block. `AppModel` must also have direct access to values stored within the application's `SharedPreferences` file via the `AppPreferences` class we created. Catering to these different demands may seem daunting at first, but is easy as pie.

The first thing we must do is add the necessary constants that will be utilized by `AppModel`. We will need to create constants for the possible game statuses and the possible motions that can occur during gameplay. These constants are created with ease with the use of `enum` classes:

```
class AppModel {
  enum class Statuses {
    AWAITING_START, ACTIVE, INACTIVE, OVER
  }

  enum class Motions {
    LEFT, RIGHT, DOWN, ROTATE
  }
}
```

We created four status constants earlier. `AWAITING_START` is the status of the game before the game has been started. `ACTIVE` is the status in which the game exists when gameplay is currently in progress. `OVER` is the status that the game takes when the game ends.

Earlier in this chapter, it was stated that four distinct motions can occur on a block. Blocks can be moved to the right, to the left, up, down, and rotated. LEFT, RIGHT, UP, DOWN, and ROTATE are defined in the Motions enum class to represent these distinct motions.

Having added the constants required, we can proceed by adding the necessary class properties of AppModel, which are as follows:

```
package com.mydomain.tetris.models

import android.graphics.Point
import com.mydomain.tetris.constants.FieldConstants
import com.mydomain.tetris.helpers.array2dOfByte
import com.mydomain.tetris.storage.AppPreferences

class AppModel {
  var score: Int = 0
  private var preferences: AppPreferences? = null

  var currentBlock: Block? = null
  var currentState: String = Statuses.AWAITING_START.name

  private var field: Array<ByteArray> = array2dOfByte(
    FieldConstants.ROW_COUNT.value,
    FieldConstants.COLUMN_COUNT.value
  )

  enum class Statuses {
    AWAITING_START, ACTIVE, INACTIVE, OVER
  }

  enum class Motions {
    LEFT, RIGHT, DOWN, ROTATE
  }
}
```

score is an integer property that will be used to hold the current score of the player within a gaming session. preferences is a private property that will hold an AppPreferences object to provide direct access to the application's SharedPreferences file. currentBlock is a property that will hold the current block translating across the play field. currentState holds the state of the game. Statuses.AWAITING_START.name returns the name of Statuses.AWAITING_START in the form of an AWAITING_START string. The current state of the game is initialized to AWAITING_START immediately because this is the first state that GameActivity must transition into upon launch. Lastly, field is a two-dimensional array that will be used as the playing field for the game.

Next we must add a few setter and getter functions. These functions are setPreferences(), setCellStatus(), and getCellStatus(). Add the following functions to AppModel:

```
fun setPreferences(preferences: AppPreferences?) {
   this.preferences = preferences
}

fun getCellStatus(row: Int, column: Int): Byte? {
   return field[row][column]
}

private fun setCellStatus(row: Int, column: Int, status: Byte?) {
   if (status != null) {
     field[row][column] = status
   }
}
```

The setPreferences() method sets the preferences property of AppModel to the AppPreferences instance passed as an argument to the function. The getCellStatus() method returns the status of a cell existing in a specified row-column position within the field's two-dimensional array. The setCellStatus() method sets the status of a cell existing in the field to a specified byte.

Functions for checking state are necessary in the model as well. These will serve as a medium to assert the state that the game is currently in. As we have three possible game statuses corresponding to three possible game states, three functions are required for each individual game state. These methods are isGameAwaitingStart(), isGameActive(), and isGameOver():

```
class AppModel {

  var score: Int = 0
  private var preferences: AppPreferences? = null

  var currentBlock: Block? = null
  var currentState: String = Statuses.AWAITING_START.name

  private var field: Array<ByteArray> = array2dOfByte(
    FieldConstants.ROW_COUNT.value,
    FieldConstants.COLUMN_COUNT.value
  )

  fun setPreferences(preferences: AppPreferences?) {
    this.preferences = preferences
  }
```

```
fun getCellStatus(row: Int, column: Int): Byte? {
  return field[row][column]
}

private fun setCellStatus(row: Int, column: Int, status: Byte?) {
  if (status != null) {
    field[row][column] = status
  }
}

fun isGameOver(): Boolean {
  return currentState == Statuses.OVER.name
}

fun isGameActive(): Boolean {
  return currentState == Statuses.ACTIVE.name
}

fun isGameAwaitingStart(): Boolean {
  return currentState == Statuses.AWAITING_START.name
}

enum class Statuses {
  AWAITING_START, ACTIVE, INACTIVE, OVER
}

enum class Motions {
  LEFT, RIGHT, DOWN, ROTATE
}
}
```

All three methods return Boolean values of either `true` or `false` depending on whether the game is existing in their respective states. So far, we have not made use of the `score` in `AppModel`. Let's add a function that can be used to increase the score value held by the score. We will name the function `boostScore()`.

```
private fun boostScore() {
  score += 10
  if (score > preferences?.getHighScore() as Int)
    preferences?.saveHighScore(score)
}
```

When called, `boostScore()` increases the current score of the player by 10 points, after which it checks whether the current score of the player is greater than the high score recorded in the preferences file. If the current score is greater than the saved high score, the high score is overwritten with the current score.

Having gotten the basic functions and fields up and running, we can progress to creating slightly more complicated functions. The first of these functions is `generateNextBlock()`:

```
private fun generateNextBlock() {
    currentBlock = Block.createBlock()
}
```

The `generateNextBlock()` function creates a new block instance and sets `currentBlock` to the newly created instance.

Before going any further with method definitions, let's create one more `enum` class to hold constant cell values. Create a `CellConstants.kt` file in the constants package and add the following source code to it:

```
package com.mydomain.tetris.constants

enum class CellConstants(val value: Byte) {
    EMPTY(0), EPHEMERAL(1)
}
```

You may be wondering what these constants are for. Recall when we created the `Frame` class to model a blocks frame, we defined `addRow()`, which took a string of 1s and 0s as its argument—with 1 representing cells that made up the frame and 0 representing cells excluded from the frame—and converted these 1s and 0s to byte representations. We are going to be manipulating these bytes in upcoming functions and we need to have corresponding constants for them.

Import the newly created `enum` class into `AppModel`. We will make use of it in the upcoming function:

```
private fun validTranslation(position: Point, shape: Array<ByteArray>):
Boolean {
    return if (position.y < 0 || position.x < 0) {
        false
    } else if (position.y + shape.size > FieldConstants.ROW_COUNT.value) {
        false
    } else if (position.x + shape[0].size > FieldConstants
                .COLUMN_COUNT.value) {
        false
    } else {
        for (i in 0 until shape.size) {
            for (j in 0 until shape[i].size) {
                val y = position.y + i
                val x = position.x + j

                if (CellConstants.EMPTY.value != shape[i][j] &&
```

```
            CellConstants.EMPTY.value != field[y][x]) {
          return false
      }
    }
  }
  true
  }
}
```

Add the preceding `validTranslation()` method to `AppModel`. As the name implies, this function is used to check whether a translational motion of a tetromino in the playing field is valid based on a set of conditions. It returns a `true` Boolean value if the translation is valid, and `false` otherwise. The first three conditionals test whether the position the tetromino is being translated in the field to is a valid one. The `else` block checks whether the cells the tetromino is attempting to translate into are empty. If they are not, `false` is returned.

We need a calling function for `validTranslation()`. We will declare `moveValid()` to serve this purpose. Add the following function to `AppModel`:

```
private fun moveValid(position: Point, frameNumber: Int?): Boolean {
  val shape: Array<ByteArray>? = currentBlock?
                            .getShape(frameNumber as Int)
  return validTranslation(position, shape as Array<ByteArray>)
}
```

`moveValid()` utilizes `validTranslation()` to check whether a move performed by the player is permitted. If the move is permitted, it returns `true`, otherwise `false` is returned. We need to create a few other important methods. These are `generateField()`, `resetField()`, `persistCellData()`, `assessField()`, `translateBlock()`, `blockAdditionPossible()`, `shiftRows()`, `startGame()`, `restartGame()`, `endGame()`, and `resetModel()`.

We will firstly work on the `generateField()` method. Add the code shown below to `AppModel`.

```
fun generateField(action: String) {
  if (isGameActive()) {
    resetField()
    var frameNumber: Int? = currentBlock?.frameNumber
    val coordinate: Point? = Point()
    coordinate?.x = currentBlock?.position?.x
    coordinate?.y = currentBlock?.position?.y

    when (action) {
```

```
            Motions.LEFT.name -> {
              coordinate?.x = currentBlock?.position?.x?.minus(1)
            }
            Motions.RIGHT.name -> {
              coordinate?.x = currentBlock?.position?.x?.plus(1)
            }
            Motions.DOWN.name -> {
              coordinate?.y = currentBlock?.position?.y?.plus(1)
            }
            Motions.ROTATE.name -> {
              frameNumber = frameNumber?.plus(1)

              if (frameNumber != null) {
                if (frameNumber >= currentBlock?.frameCount as Int) {
                  frameNumber = 0
                }
              }
            }
          }

          if (!moveValid(coordinate as Point, frameNumber)) {
            translateBlock(currentBlock?.position as Point,
                          currentBlock?.frameNumber as Int)
            if (Motions.DOWN.name == action) {
              boostScore()
              persistCellData()
              assessField()
              generateNextBlock()

              if (!blockAdditionPossible()) {
                currentState = Statuses.OVER.name;
                currentBlock = null;
                resetField(false);
              }
            }
          } else {
            if (frameNumber != null) {
              translateBlock(coordinate, frameNumber)
              currentBlock?.setState(frameNumber, coordinate)
            }
          }
        }
      }
    }
```

generateField() generates a refresh of the field. This field refresh is determined by the action that is passed as the argument of generateField().

First, generateField() checks whether the game is currently in its active state when called. If the game is active, the frame number and coordinates of the block are retrieved. The action requested is then determined via the when expression. Having determined the requested action, the coordinates of the block are changed appropriately if the action requested is a leftward, rightward, or downward motion. If a rotational motion is requested, frameNumber is changed to an appropriate number of a frame that represents the tetromino in the rotation exerted.

The generateField() method then checks whether the motion requested is a valid motion via moveValid(). If the move is not valid, the current block is fixed in the field to its current position with the use of translateBlock().

The resetField(), persistCellData() and assessField() methods invoked by generateField() are given below. Add them to AppModel:

```
private fun resetField(ephemeralCellsOnly: Boolean = true) {
    for (i in 0 until FieldConstants.ROW_COUNT.value) {
        (0 until FieldConstants.COLUMN_COUNT.value)
            .filter { !ephemeralCellsOnly || field[i][it] ==
                    CellConstants.EPHEMERAL.value }
            .forEach { field[i][it] = CellConstants.EMPTY.value }
    }
}

private fun persistCellData() {
    for (i in 0 until field.size) {
        for (j in 0 until field[i].size) {
            var status = getCellStatus(i, j)

            if (status == CellConstants.EPHEMERAL.value) {
                status = currentBlock?.staticValue
                setCellStatus(i, j, status)
            }
        }
    }
}

private fun assessField() {
    for (i in 0 until field.size) {
        var emptyCells = 0;

        for (j in 0 until field[i].size) {
```

```
      val status = getCellStatus(i, j)
      val isEmpty = CellConstants.EMPTY.value == status
      if (isEmpty)
        emptyCells++
  }
  if (emptyCells == 0)
    shiftRows(i)
}
}
```

As you may have noticed, translateBlock() has not been implemented. Go ahead and add this method along with blockAdditionPossible(), shiftRows(), startGame(), restartGame(), endGame(), and resetModel() to AppModel is as follows:

```
private fun translateBlock(position: Point, frameNumber: Int) {
  synchronized(field) {
    val shape: Array<ByteArray>? = currentBlock?.getShape(frameNumber)

    if (shape != null) {
      for (i in shape.indices) {
        for (j in 0 until shape[i].size) {
          val y = position.y + i
          val x = position.x + j

          if (CellConstants.EMPTY.value != shape[i][j]) {
            field[y][x] = shape[i][j]
          }
        }
      }
    }
  }
}

private fun blockAdditionPossible(): Boolean {
  if (!moveValid(currentBlock?.position as Point,
      currentBlock?.frameNumber)) {
    return false
  }
  return true
}

private fun shiftRows(nToRow: Int) {
  if (nToRow > 0) {
    for (j in nToRow - 1 downTo 0) {
      for (m in 0 until field[j].size) {
        setCellStatus(j + 1, m, getCellStatus(j, m))
      }
    }
  }
}
```

```
      }
   }

   for (j in 0 until field[0].size) {
      setCellStatus(0, j, CellConstants.EMPTY.value)
   }
}

fun startGame() {
   if (!isGameActive()) {
      currentState = Statuses.ACTIVE.name
      generateNextBlock()
   }
}

fun restartGame() {
   resetModel()
   startGame()
}

fun endGame() {
   score = 0
   currentState = AppModel.Statuses.OVER.name
}

private fun resetModel() {
   resetField(false)
   currentState = Statuses.AWAITING_START.name
   score = 0
}
```

In a scenario where the requested move is a downward motion and the move is not valid, it implies that the block has reached the bottom of the field. In this case, the player's score is boosted via `boostScore()` and the states of all cells in the field are persisted via `persistCellData()`. The `assessField()` method is then called to scan through the field row by row and check whether all cells in a row have been filled up:

```
private fun assessField() {
   for (i in 0 until field.size) {
      var emptyCells = 0;

      for (j in 0 until field[i].size) {
         val status = getCellStatus(i, j)
         val isEmpty = CellConstants.EMPTY.value == status

         if (isEmpty)
            emptyCells++
```

```
        }

        if (emptyCells == 0)
            shiftRows(i)
    }
}
```

In the case where all cells in a row are filled up, the row is cleared and shifted by `shiftRow()`. After the assessment of the field is complete, a new block is generated with `generateNextBlock()`:

```
private fun generateNextBlock() {
    currentBlock = Block.createBlock()
}
```

Before the newly generated block can be pushed to the field, `AppModel` makes sure that the field is not already filled up and the block can be moved into the field with `blockAdditionPossible()`:

```
private fun blockAdditionPossible(): Boolean {
    if (!moveValid(currentBlock?.position as Point,
        currentBlock?.frameNumber)) {
        return false
    }
    return true
}
```

If block addition is not possible, that means all blocks have been stacked to the top edge of the field. This results in a game over. As a result, the current state of the game is set to `Statuses.OVER` and the `currentBlock` is set to `null`. Lastly, the field is cleared.

On the other hand, if the move was valid from the start, the block is translated to its new coordinates via `translateBlock()` and the state of the current block is set to its new coordinates and `frameNumber`.

With those additions in place, we have been able to successfully create the application model to handle the gameplay logic. Now we have to create a view that exploits `AppModel`.

Creating TetrisView

So far, so good. We have successfully implemented classes to model blocks, frames, and shapes of different tetrominoes that will be used within the application, as well as implemented an `AppModel` class to coordinate all the interactions between views and these programmatic components created. Without this view existing, there is no means by which a user can interact with `AppModel`. If a user cannot interact with the game, the game might as well not exist. In this section, we will implement `TetrisView`, the user interface by which a user will play Tetris.

Create a package named `view` in your source package and add a `TetrisView.kt` file in it. As we want `TestrisView` to be a `View`, we must declare it to extend the View class. Add the code below to `TetrisView.kt`.

```kotlin
package com.mydomain.tetris.views

import android.content.Context
import android.graphics.Canvas
import android.graphics.Color
import android.graphics.Paint
import android.graphics.RectF
import android.os.Handler
import android.os.Message
import android.util.AttributeSet
import android.view.View
import android.widget.Toast
import com.mydomain.tetris.constants.CellConstants
import com.mydomain.tetris.GameActivity
import com.mydomain.tetris.constants.FieldConstants
import com.mydomain.tetris.models.AppModel
import com.mydomain.tetris.models.Block

class TetrisView : View {

    private val paint = Paint()
    private var lastMove: Long = 0
    private var model: AppModel? = null
    private var activity: GameActivity? = null
    private val viewHandler = ViewHandler(this)
    private var cellSize: Dimension = Dimension(0, 0)
    private var frameOffset: Dimension = Dimension(0, 0)

    constructor(context: Context, attrs: AttributeSet) :
            super(context, attrs)

    constructor(context: Context, attrs: AttributeSet, defStyle: Int) :
```

```
                    super(context, attrs, defStyle)

    companion object {
      private val DELAY = 500
      private val BLOCK_OFFSET = 2
      private val FRAME_OFFSET_BASE = 10
    }
  }
```

The `TetrisView` class is declared to extend `View`. `View` is a class that all application view elements must extend. As the `View` type has a constructor that must be initialized, we are declaring two secondary constructors for `TetrisView` that initialize two distinct constructors of the view class, depending on which secondary constructor is called.

The `paint` property is an instance of `android.graphics.Paint`. The `Paint` class holds style and color information concerning how to draw texts, bitmaps, and geometries. `lastMove` will be used to keep track of the last time in milliseconds that a move was made. The `model` instance will be used to hold an `AppModel` instance that will be interacted with by `TetrisView` to control gameplay. Activity is an instance of the `GameActivity` class we created. The `cellSize` and `frameOffset` are properties that will hold dimensions for the size of cells in the game and the frame offset, respectively.

Neither `ViewHandler` nor `Dimension` is a class provided to us by the Android application framework. We must implement these two classes.

Implementing ViewHandler

As blocks will be moving along the fields in intervals with a constant time delay, we need a means of putting the thread that handles the movement of blocks to sleep and waking the thread to make a block motion after a period of time. A good way to take care of this requirement is to use a handler to process message delay requests and continue message handling after the delay has completed. Putting this in more direct terms, according to Android's documentation, *the handler allows you to send and process Message objects associated with a thread's MessageQueue*. Every handler instance is associated with a thread and the thread's message queue.

`ViewHandler` is a custom handler we will implement for `TetrisView` that caters to the view's message-sending and processing needs. As `ViewHandler` is subclass of `Handler`, we must extend `Handler` and add our necessary behavior to the `ViewHandler` class.

Add the following `VieHandler` class as a private class within `TetrisView`:

```
private class ViewHandler(private val owner: TetrisView) : Handler() {

  override fun handleMessage(message: Message) {
    if (message.what == 0) {
      if (owner.model != null) {
        if (owner.model!!.isGameOver()) {
          owner.model?.endGame()
          Toast.makeText(owner.activity, "Game over",
                      Toast.LENGTH_LONG).show();
        }
        if (owner.model!!.isGameActive()) {
          owner.setGameCommandWithDelay(AppModel.Motions.DOWN)
        }
      }
    }
  }

  fun sleep(delay: Long) {
    this.removeMessages(0)
    sendMessageDelayed(obtainMessage(0), delay)
  }
}
```

The `ViewHandler` class takes an instance of `TetrisView` as an argument in its constructor. `ViewHandler` overrides the `handleMessage()` function existing in its superclass class. `handleMessage()` checks that the what message was sent. The `what` is an integer value denoting the message sent. If `what` is equal to 0, and the instance—owner—of `TetrisView` passed possesses a model that is not equal to 0, some statuses of the game are checked. If the game is over, it will call `endGame()` of `AppModel` function and show a popup alerting the player that the game is over. If the game is in its active state, a down motion is fired.

The `sleep()` method simply removes any previously sent message and sends a new message with a delay specified by the delay argument.

Implementing Dimension

`Dimension` only needs to be able to hold two properties: width and height. As such, it is a perfect candidate for the utilization of a data class. Add the following private class to the `TetrisView` class:

```
private data class Dimension(val width: Int, val height: Int)
```

The preceding one-liner provides us with the properties, as well as the setter and getters we need for them.

Implementing TetrisView

As you may have guessed, at this point `TetrisView` is far from completion. First and foremost we must implement a few setter methods for the `model` and `activity` properties of the view. These methods are shown below. Make sure to add them to your `TetrisView` class.

```
fun setModel(model: AppModel) {
    this.model = model
}

fun setActivity(gameActivity: GameActivity) {
    this.activity = gameActivity
}
```

`setModel()` and `setActivity()` are setter functions for the model and activity instance properties. As the names imply, `setModel()` sets the current model in use by the view and `setActivity()` sets the activity in use. Now, let us add three additional methods `setGameCommand()`, `setGameCommandWithDelay()` and `updateScore()`.

```
fun setGameCommand(move: AppModel.Motions) {
    if (null != model && (model?.currentState ==
AppModel.Statuses.ACTIVE.name)) {
        if (AppModel.Motions.DOWN == move) {
            model?.generateField(move.name)
            invalidate()
            return
        }
        setGameCommandWithDelay(move)
    }
}

fun setGameCommandWithDelay(move: AppModel.Motions) {
    val now = System.currentTimeMillis()

    if (now - lastMove > DELAY) {
        model?.generateField(move.name)
        invalidate()
        lastMove = now
    }
    updateScores()
    viewHandler.sleep(DELAY.toLong())
```

```
}

private fun updateScores() {
  activity?.tvCurrentScore?.text = "${model?.score}"
  activity?.tvHighScore?.text =
"${activity?.appPreferences?.getHighScore()}"
}
```

setGameCommand() sets the current motion command being executed by the game. If a DOWN motion command is in execution, the application model generates the field for a block experiencing a downward motion. The invalidate() method being called within setGameCommand() can be taken as a request to draw a change on the screen. invalidate() ultimately results in a call to onDraw().

onDraw() is a method that is inherited from the View class. It is called when a view should render its content. We will need to provide a custom implementation of this for our view. Add the code below to your TetrisView class.

```
override fun onDraw(canvas: Canvas) {
  super.onDraw(canvas)
  drawFrame(canvas)

  if (model != null) {
    for (i in 0 until FieldConstants.ROW_COUNT.value) {
      for (j in 0 until FieldConstants.COLUMN_COUNT.value) {
        drawCell(canvas, i, j)
      }
    }
  }
}

private fun drawFrame(canvas: Canvas) {
  paint.color = Color.LTGRAY

  canvas.drawRect(frameOffset.width.toFloat(),
          frameOffset.height.toFloat(), width -
frameOffset.width.toFloat(),
          height - frameOffset.height.toFloat(), paint)
}

private fun drawCell(canvas: Canvas, row: Int, col: Int) {
  val cellStatus = model?.getCellStatus(row, col)

  if (CellConstants.EMPTY.value != cellStatus) {
    val color = if (CellConstants.EPHEMERAL.value == cellStatus) {
      model?.currentBlock?.color
```

```
    } else {
      Block.getColor(cellStatus as Byte)
    }
    drawCell(canvas, col, row, color as Int)
  }
}

private fun drawCell(canvas: Canvas, x: Int, y: Int, rgbColor: Int) {
  paint.color = rgbColor

  val top: Float = (frameOffset.height + y * cellSize.height +
BLOCK_OFFSET).toFloat()
  val left: Float = (frameOffset.width + x * cellSize.width +
BLOCK_OFFSET).toFloat()
  val bottom: Float = (frameOffset.height + (y + 1) * cellSize.height -
BLOCK_OFFSET).toFloat()
  val right: Float = (frameOffset.width + (x + 1) * cellSize.width -
BLOCK_OFFSET).toFloat()
  val rectangle = RectF(left, top, right, bottom)

  canvas.drawRoundRect(rectangle, 4F, 4F, paint)
}

override fun onSizeChanged(width: Int, height: Int, previousWidth: Int,
previousHeight: Int) {
  super.onSizeChanged(width, height, previousWidth, previousHeight)

  val cellWidth = (width - 2 * FRAME_OFFSET_BASE) /
FieldConstants.COLUMN_COUNT.value
  val cellHeight = (height - 2 * FRAME_OFFSET_BASE) /
FieldConstants.ROW_COUNT.value
  val n = Math.min(cellWidth, cellHeight)
  this.cellSize = Dimension(n, n)
  val offsetX = (width - FieldConstants.COLUMN_COUNT.value * n) / 2
  val offsetY = (height - FieldConstants.ROW_COUNT.value * n) / 2
  this.frameOffset = Dimension(offsetX, offsetY)
}
```

The onDraw() method in TetrisView overrides the onDraw() in its superclass. onDraw(), takes a canvas object as its only argument and must call the onDraw() function in its superclass. This is done by invoking super.onDraw() and passing the canvas instance as an argument.

After invoking `super.onDraw()`, `onDraw()` in `TetrisView` invokes `drawFrame()`, which draws the frame for `TetrisView`. After which, individual cells are drawn within the canvas by utilizing the `drawCell()` functions we created.

The `setGameCommandWithDelay()` works similarly to `setGameCommand()` with the exception that updates the game score and it puts `viewHandler` to sleep after executing the game command. The `updateScore()` function is used to update the current score and high score text views in game activity.

The `onSizeChanged()` is a function that is called when the size of a view has changed. The function provides access to the current width and height of the view, as well as its former width and height. As with other overriden functions we have used, we invoke its counterpart function in its super class. We use the width and height arguments provided to us to calculate and set dimensions for the size of each cell—`cellSize`. Finally, in `onSizeChanged()`, the `offsetX` and `offsetY` are calculated and used to set `frameOffset`.

Finishing up GameActivity

So far, you have successfully implemented the views, handlers, helper functions, classes, and models necessary to put the Tetris game together. Now we are going to finish up the work we started by putting it all together in `GameActivity`. The first thing on our agenda is to add the newly created `tetris` view to the game activity's layout. We can easily add `TetrisView` as a child element anywhere within a layout file by utilizing the `<com.mydomain.tetris.views.TetrisView>` layout tag:

```xml
<?xml version="1.0" encoding="utf-8"?>
<android.support.constraint.ConstraintLayout
xmlns:android="http://schemas.android.com/apk/res/android"
    xmlns:app="http://schemas.android.com/apk/res-auto"
    xmlns:tools="http://schemas.android.com/tools"
    android:layout_width="match_parent"
    android:layout_height="match_parent"
    tools:context="com.mydomain.tetris.GameActivity">
  <LinearLayout
        android:layout_width="match_parent"
        android:layout_height="match_parent"
        android:orientation="horizontal"
        android:weightSum="10"
        android:background="#e8e8e8">
    <LinearLayout
        android:layout_width="wrap_content"
```

```xml
        android:layout_height="match_parent"
        android:orientation="vertical"
        android:gravity="center"
        android:paddingTop="32dp"
        android:paddingBottom="32dp"
        android:layout_weight="1">
    <LinearLayout
            android:layout_width="wrap_content"
            android:layout_height="0dp"
            android:layout_weight="1"
            android:orientation="vertical"
            android:gravity="center">
        <TextView
            android:layout_width="wrap_content"
            android:layout_height="wrap_content"
            android:text="@string/current_score"
            android:textAllCaps="true"
            android:textStyle="bold"
            android:textSize="14sp"/>
        <TextView
            android:id="@+id/tv_current_score"
            android:layout_width="wrap_content"
            android:layout_height="wrap_content"
            android:textSize="18sp"/>
        <TextView
            android:layout_width="wrap_content"
            android:layout_height="wrap_content"
            android:layout_marginTop="@dimen/layout_margin_top"
            android:text="@string/high_score"
            android:textAllCaps="true"
            android:textStyle="bold"
            android:textSize="14sp"/>
        <TextView
            android:id="@+id/tv_high_score"
            android:layout_width="wrap_content"
            android:layout_height="wrap_content"
            android:textSize="18sp"/>
    </LinearLayout>
    <Button
            android:id="@+id/btn_restart"
            android:layout_width="wrap_content"
            android:layout_height="wrap_content"
            android:text="@string/btn_restart"/>
</LinearLayout>
<View
        android:layout_width="1dp"
        android:layout_height="match_parent"
        android:background="#000"/>
```

```
<LinearLayout
        android:layout_width="0dp"
        android:layout_height="match_parent"
        android:layout_weight="9">
  <!-- Adding TetrisView -->
  <com.mydomain.tetris.views.TetrisView
          android:id="@+id/view_tetris"
          android:layout_width="match_parent"
          android:layout_height="match_parent" />

  </LinearLayout>
 </LinearLayout>
</android.support.constraint.ConstraintLayout>
```

Once you have added the `tetris` view to `activity_game.xml`, open the `GameActivity` class and employ the changes to the class shown in the following code block:

```
package com.mydomain.tetris

import android.os.Bundle
import android.support.v7.app.AppCompatActivity
import android.view.MotionEvent
import android.view.View
import android.widget.Button
import android.widget.TextView
import com.mydomain.tetris.models.AppModel
import com.mydomain.tetris.storage.AppPreferences
import com.mydomain.tetris.views.TetrisView

class GameActivity: AppCompatActivity() {

  var tvHighScore: TextView? = null
  var tvCurrentScore: TextView? = null
  private lateinit var tetrisView: TetrisView

  var appPreferences: AppPreferences? = null
  private val appModel: AppModel = AppModel()

  public override fun onCreate(savedInstanceState: Bundle?) {
    super.onCreate(savedInstanceState)
    setContentView(R.layout.activity_game)
    appPreferences = AppPreferences(this)
    appModel.setPreferences(appPreferences)

    val btnRestart = findViewById<Button>(R.id.btn_restart)
    tvHighScore = findViewById<TextView>(R.id.tv_high_score)
    tvCurrentScore = findViewById<TextView>(R.id.tv_current_score)
```

```kotlin
    tetrisView = findViewById<TetrisView>(R.id.view_tetris)
    tetrisView.setActivity(this)
    tetrisView.setModel(appModel)

    tetrisView.setOnTouchListener(this::onTetrisViewTouch)
    btnRestart.setOnClickListener(this::btnRestartClick)

    updateHighScore()
    updateCurrentScore()
}

private fun btnRestartClick(view: View) {
    appModel.restartGame()
}

private fun onTetrisViewTouch(view: View, event: MotionEvent):
            Boolean {
    if (appModel.isGameOver() || appModel.isGameAwaitingStart()) {
        appModel.startGame()
        tetrisView.setGameCommandWithDelay(AppModel.Motions.DOWN)

    } else if(appModel.isGameActive()) {
        when (resolveTouchDirection(view, event)) {
            0 -> moveTetromino(AppModel.Motions.LEFT)
            1 -> moveTetromino(AppModel.Motions.ROTATE)
            2 -> moveTetromino(AppModel.Motions.DOWN)
            3 -> moveTetromino(AppModel.Motions.RIGHT)
        }
    }
    return true
}

private fun resolveTouchDirection(view: View, event: MotionEvent):
            Int {
    val x = event.x / view.width
    val y = event.y / view.height
    val direction: Int

    direction = if (y > x) {
        if (x > 1 - y) 2 else 0
    }
    else {
        if (x > 1 - y) 3 else 1
    }
    return direction
}

private fun moveTetromino(motion: AppModel.Motions) {
```

```
    if (appModel.isGameActive()) {
      tetrisView.setGameCommand(motion)
    }
  }

  private fun updateHighScore() {
    tvHighScore?.text = "${appPreferences?.getHighScore()}"
  }

  private fun updateCurrentScore() {
    tvCurrentScore?.text = "0"
  }
}
```

We added an object reference to the tetris view layout element in activity_game.xml in the form of the tetrisView property; we also created an instance of AppModel that will be used by GameActivity. In oncreate(), we set the activity in use by tetrisView to the current instance of the GameActivity and set the model in use by tetrisView to appModel – the AppModel instance property we created. In addition, the on-touch listener for tetrisView was set to the onTetrisViewTouch() function.

If tetrisView is touched and the game is in an AWAITING_START or OVER state, a new game is started. If tetrisView is touched and the game is in its ACTIVE state, the direction in which the touch on tetrisView occurred is resolved with the help of resolveTouchDirection(). moveTetromino() is used to move a tetromino block based on the action passed to it. If a left touch occurred, moveTetromino() is called with AppModel.Motions.LEFT set as its argument. This results in the movement of the tetromino to the left on the field. Right, down, and up touches on tetrisView result in rightward, downward, and rotational motions.

Having made all the additions, build and run the project. Once the project launches on your desired device, navigate to game activity and touch the `tetris` view to the right of the screen. The game will start:

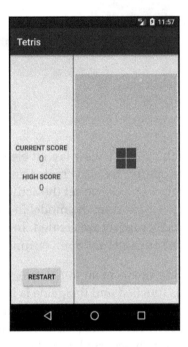

Feel free to play around with the game you created. You deserve it!

Introduction to Model-View-Presenter (MVP)

Over the course of developing the Tetris application, we attempted to add structure across our code base by separating out program files into different packages based on the tasks they performed. We tried to abstract application logic into the `AppModel` class, and user interactions related to gameplay to be handled by the `TetrisView` view class. This certainly brought some order into our code base in contrast with, say, putting all logic into one big class file.

Needless to say, there are better ways to separate concerns within an Android application. One way is the MVP pattern.

What is MVP?

MVP is a common pattern in Android that is derived from the **Model-View-Controller** (**MVC**) pattern. MVP attempts to view related concerns from application logic. There are many reasons for which this is done, such as:

- To increase the maintainability of a code base
- To improve application reliability

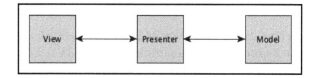

 Let's familiarize ourselves with the actors in the MVP pattern.

Model

In MVP, models are interfaces that have the task of managing data. The responsibilities of models include interacting with databases, making API calls, communicating over networks, and coordinating objects and other programmatic components to perform specific tasks.

View

Views are application entities that display content to users and serve as an interface for user input. A view can be an activity, a fragment, or an Android widget. A view is responsible for rendering data in a way decided upon by the presenter.

Presenter

A presenter is a layer that acts as a middleman between a view and a model. The major responsibility of the presenter is querying the model and updating a view. Put simply, presentation logic goes into the presenter. An important thing to keep in mind is that a presenter has a one-to-one relationship with a view.

Varying implementations of MVP

The MVP pattern has varying means by which it is implemented in practice. For example, some implementations of MVP utilize a *contract* to describe the interaction between the view and presenter.

In addition, there are implementations of MVP that utilize lifecycle callbacks within the presenter, such as onCreate(). This is in an attempt to mirror callbacks existing in the activity lifecycle. Other implementations discard the implementation of these callbacks entirely.

In reality, there is no one true implementation of MVP in Android applications, but there are best practices that can be followed while implementing MVP. You will learn about these best practices and have hands-on experience with developing an MVP application with Kotlin in Chapter 5, *Building the Messenger Android App*.

Summary

In this chapter, we got down and dirty with Kotlin by implementing a classic game, *Tetris*. Over the course of this chapter, we learned about a vast array of things, such as how to model logical components of an application with classes, access and visibility modifiers, how to create views and handlers in Android applications, the utilization of data classes to easily create data models, and the MVP pattern.

In the next chapter, we will apply our knowledge of Kotlin to the web domain by implementing the backend of a messenger application.

4
Designing and Implementing the Messenger Backend with Spring Boot 2.0

In the last couple of chapters, we gained a firm grasp of the fundamentals of the Kotlin programming language by implementing the classic game, *Tetris*. In `Chapter 3`, *Implementing Tetris Logic and Functionality*, we finished the development of the game by implementing its application logic. We created programmatic models for blocks, shapes, frames, and the application as a whole, via an app model class. In addition, we learned how to create custom views by implementing Tetris view—a view that the user of the application interacted with to play the game.

We will hone our Kotlin development skills further by developing a simple messenger application for the Android platform. In the process of implementing the Android application, we will first develop a RESTful API that will provide web content to the application behind the scenes. The application programming interface will be built with Spring Boot 2.0. After developing the application programming interface, we will deploy it to a remote server. Over the course of this chapter, you will learn about the following topics:

- Basic system design
- Modeling system behavior with state diagrams
- Database design fundamentals
- Modeling a database with entity relationship (E-R) diagrams

- Building backend microservices with Spring Boot 2.0
- Working with PostgreSQL
- Dependency management with Maven
- Amazon Web Services (AWS)

Without further ado, let's dive into this chapter by designing the messenger application programming interface.

Designing the messenger API

To design a fully functional RESTful application programming interface for our messenger android application, we must fully understand the concept of application programming interfaces, **Representational State Transfer** (**REST**) and RESTful services.

Application programming interfaces

An application programming interface is a collection of functions, routines, procedures, protocols, and resources that can be used for building software. In other words, an application programming interface—API for short—is a collection of well-defined and appropriately structured methods or channels of communication between software components.

Application programming interfaces can be developed for use with various application domains. Some common application domains that APIs are developed for include the development of web-based systems, operating systems, and computer hardware, as well as interaction with embedded systems.

REST

Restful state transfer is a way of facilitating functional operation and interactions between two or more distinct systems (or subsystems) via the internet. Web services that adhere to REST allow interacting systems to access web content; they also perform authorized operations on web content that they have access to. These inter-system communications are done using a well-defined set of stateless operations. A RESTful web service adheres to REST and provides web content to communicating systems via predetermined stateless operations.

In the present day, numerous systems that communicate with web services utilize REST. Systems that may utilize REST are based on the client-server architecture. The API we will be developing is based on REST, and, as such, will make use of representational state transfer.

Designing the messenger API system

In this section, we will attempt to concisely design the messenger API system. You may be wondering at this point what exactly system design is and what it entails. Fear not, we will explain these in the next few paragraphs.

System design is the process of defining the architecture, modules, interfaces, and data for a system to satisfy specified requirements that are from a pre-performed system analysis phase. System design consists of numerous processes and the utilization of different design orientations. In addition, the in-depth designing of systems requires the understanding of numerous topics, such as coupling and cohesion, which are far beyond the scope of this book. With this in mind, we shall attempt to give basic definitions of the interactions and data utilized in our system. We shall achieve this by designing the system incrementally.

Incremental development

Incremental development is an approach that can be used for the development of systems. Incremental development utilizes the incremental build model. The incremental build model is a method of developing software in which a product is designed, implemented, and tested incrementally. We will be developing the messenger API incrementally. We will in no way attempt to specify everything needed by the messenger API before we start coding. We will determine a set of specifications to get us going with the development and then create some functionality, after which we will repeat the process.

To comfortably utilize the incremental development methodology, we must utilize software that does not penalize us for making changes over the development process, such as a case where we need to change the type of data catered for by the system. Spring Boot is a perfect candidate for developing systems incrementally as it enables quick and easy changes to systems.

Up to this point, we have referred to Spring Boot a couple of times but have neither discussed what it is nor what it is used for. Let's use this opportunity to quickly do that.

Spring Boot

Spring Boot is a web application framework that was designed and developed for the purpose of the boot strapping and development of Spring applications. Spring is a web application framework that facilitates the development of web applications for the Java platform. Spring Boot makes creating industrial-strength production-grade spring-based applications easy.

We will explore how to create web applications with Spring Boot over the course of this chapter, but now is not the time for that. Before we start developing the application, we must specify what the application actually does (we cannot build something when we don't know how it works, after all).

What the messenger system does

Here, we will determine the initial requirements of the messenger system and what activities can occur within the system. We will identify high-level use cases of the messenger application.

Use cases

A use case is a statement about how an entity uses a system. Entity here refers to a type of user or a component interacting with the system. In use case definitions, entities can also be referred to as actors.

We start off by identifying the actors in the messenger system. A glaringly obvious actor is the user of the application (the person that uses the application to satisfy their messaging needs). Another actor that ideally should be considered is the admin. For the purpose of this simple messenger application though, we will cater for the single user actor. The use cases of the user are as follows:

- The user uses the messenger platform to send and receive messages
- The user uses the messenger platform to view other users on the messenger app
- The user uses the messenger platform to set and update his status
- The user can sign up to the messenger platform
- The user can log in to the messenger platform

The preceding use cases are enough to get us going. If, at any point during the course of system development, we come across a new use case, we can easily add that to the system. Now that we have identified the use cases of the system, we must properly describe the behavior of the system in catering for these use cases.

System behavior

We define system behavior to have an accurate idea of what the system does, as well as to clearly describe the interactions between components of the system. As this is a very simple application, we can clearly describe application behavior with the aid of a diagram. We will make use of a state diagram to properly describe this behavior.

State diagrams: A state diagram is used to describe the behavior of systems. A state diagram describes a system utilizing different possible states. In a state diagram, there exists a finite number of possible states that a system can be in.

The following is the state diagram for our system, taking into consideration the defined use cases:

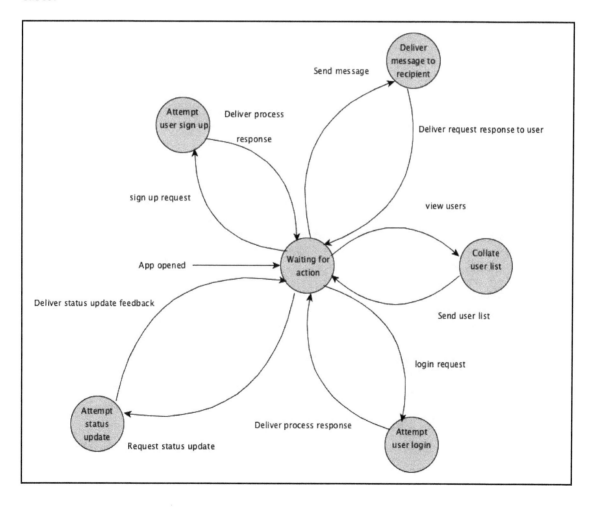

Each circle in the preceding diagram represents a state of execution of the system at a point in time. Each arrow represents an action a user can request to be carried out by the system. Upon initial start, the API waits for requests from client applications. This behavior is shown in the **Waiting for action** state. When an action request is received by the API from the client application, the system goes out of the **Waiting for action** state and services the request sent with an appropriate process.

For example, when a user requests a status update from the Android application, the server leaves the **Waiting for action** state and executes the **Attempt status update** process, after which it goes back to the **Waiting for action** state.

Identifying data

It is important to have an idea of the type of data needed before implementing a system. We can easily identify this data from the use case definitions we gave earlier. From our use case specifications, we can determine that two basic types of data are required. These are the user data and the message data. As the names imply, the user data is the data required of every user, and message data is the data pertaining to a message sent. We are not concerned with things such as schemas, entities, or entity relationship diagrams yet. We just require an idea of the data to be required by the system.

As this is a messenger app, the user we will require a username, phone number, password, and a status message. It will also be useful to keep track of the status of their account to know whether the account of a particular user is activated or has been deactivated for some reason. Not much is required as of now pertaining the messages sent. We need to keep track of the sender of a message and the intended recipient of the message.

That is all for now with respect to the data needed. We will identify more data required as the development of the application progresses, but, for now, let's code.

Implementing the messenger backend

Now that we have some sense of direction pertaining to the use cases of the messenger system as a whole, the data required within the system, and the behavior of the system, we can get started with developing the backend of the system. As we have said earlier, we will be utilizing Spring Boot to develop the messenger API because it is a perfect candidate for incremental development. In addition to this, Kotlin and Spring Boot function very well together.

As we will be handling data within the messenger API, we will need a suitable database to store the data needed by the messenger system. We will making use of PostgreSQL as our database. Let's take a brief look at PostgreSQL.

PostgreSQL

PostgreSQL is an object-relational database management system that puts particular emphasis on extensibility and standards compliance. PostgreSQL is known as Postgres. It is commonly utilized as a database server. When utilized in this way, its primary functions are to securely store data and return the data stored upon request by software applications.

There are numerous advantages to using PostgreSQL as a datastore. Some of these advantages are:

- **Extensibility**: The features of PostgreSQL can be easily and readily extended by its users. This is because its source code is available to all for free.
- **Portability**: PostgreSQL is available for all major platforms. Versions of PostgreSQL are available for almost every UNIX brand. Windows compatibility is also made possible via they Cygwin framework.
- **Integrity**: Ready availability of GUI-based tools that facilitate easy interactions with PostgreSQL.

Installing PostgreSQL

The installation of PostgreSQL is straightforward on all platforms. This section highlights its installation process on Windows, macOS, and Linux.

Windows installation

To install PostgreSQL on Windows, perform the following steps:

1. Download and run an appropriate version of the Windows PostgreSQL interactive installer. This can be downloaded from
   `https://www.enterprisedb.com/downloads/postgres-postgresql-downloads#windows`.
2. Install PostgreSQL as a Windows service. Make sure you take note of the PostgreSQL Windows service account name and password. You will need these details later in the installation process.

3. Select the **PL/pgsql** procedural language to be installed when prompted by the installer to do so.
4. You may choose to install **pgAdmin** when directed to the **Installation options** screen. If you install **pgAdmin**, enable the **Adminpack** contrib module when prompted by the installer.

If you follow the previous steps properly, PostgreSQL will be successfully installed on your system.

macOS installation

PostgreSQL can be easily installed on macOS with Homebrew. If you do not already have Homebrew installed on your system, consult Chapter 1, *The Fundamentals* for its installation instructions. Once you have determined that Homebrew is installed on your system, open your terminal and run the following command:

```
brew search postgres
```

Follow the installation instructions when prompted in the Terminal. Along the course of the installation, you may be asked for the admin password of your system. Input the password and wait for the installation to finish. You will be prompted once the installation is complete.

Linux installation

PostgreSQL can easily be installed on Linux with the use of the PostgreSQL Linux installer:

1. Go to the PostgreSQL installer download web page
   at https://www.enterprisedb.com/downloads/postgres-postgresql-downloads.
2. Select the PostgreSQL version you want to install.
3. Select an appropriate Linux installer for PostgreSQL.
4. Click the download button to download the installer.
5. Run the installer once it has been downloaded and follow all installation instructions appropriately.
6. Once you have provided the information needed by the installer, PostgreSQL will be installed on your system.

Now that we have set up PostgreSQL on our system, we can commence with the creation of the messenger API.

Creating a new Spring Boot application

The initial creation on a Spring Boot application is easy with the utilization of IntelliJ IDE and the Spring initializer. Open IntelliJ IDE and create a new project with the Spring Initializer. This can be done by clicking on **Create New Project** and selecting **Spring Initializer** on the left side bar of the **New Project** screen:

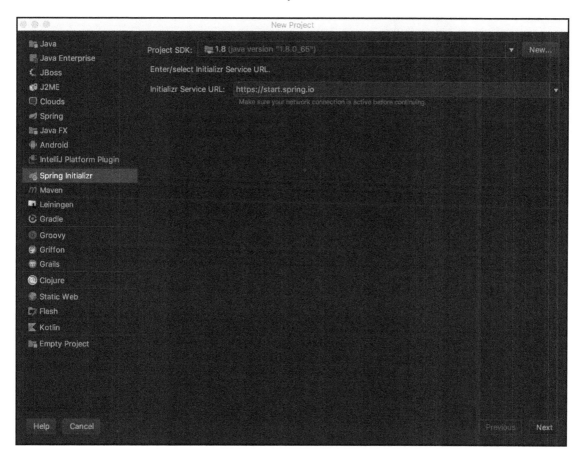

After selecting the Spring Initializer, progress to the next screen by clicking the **Next** button. Upon clicking next, before the next screen is displayed, Spring Initializer will be retrieved by the IDE. This will only take a few moments to do.

 The Spring plugin is only available in the Ultimate edition of IntelliJ IDEA, which comes with a paid subscription.

Once the Spring Initializer is retrieved, you will be asked to provide the appropriate details for the project to be created. Fill in the necessary details. You may choose to use the details used for the development of the application in this book or decide to input your own. However, in the instance that you want to use our details, do the following:

1. Input `com.example` as the group ID.
2. Enter `messenger-api` as the artifact ID.
3. Select **Maven Project** as the project type if it is not already selected.
4. Leave the packaging option and the Java version the way they are.
5. Select **Kotlin** as the language. This is important, as we are learning the Kotlin language, after all.
6. Leave the **SNAPSHOT** value as it is.
7. Enter a description of your choice.
8. Input `com.example.messenger.api` as the package name.

After filling in the required project information, proceed to the next screen by clicking **Next**:

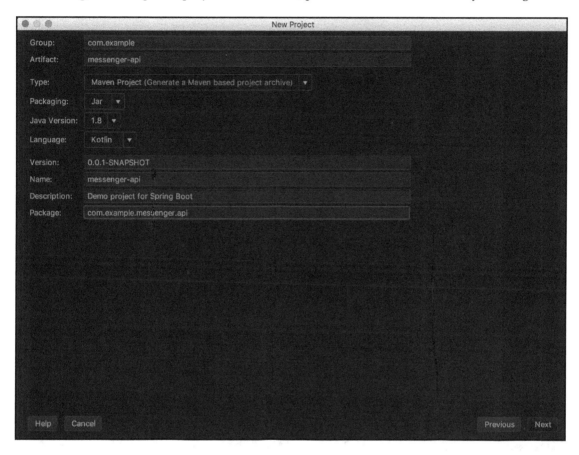

In the next screen, you will be asked to select the project dependencies. For starters, we will need to select the Security, Web, JPA, and PostgreSQL dependencies. **Security** can be found under the **Core** category, **Web** can be found under the **Web** category, and **JPA** and **PostgreSQL** can be found under the **SQL** category. In addition, in the **Spring Boot Version** selection drop-down menu at the top of the screen, select **2.0.0 M5** as the version.

After the selection of the necessary dependencies, the content should look similar to the following screenshot:

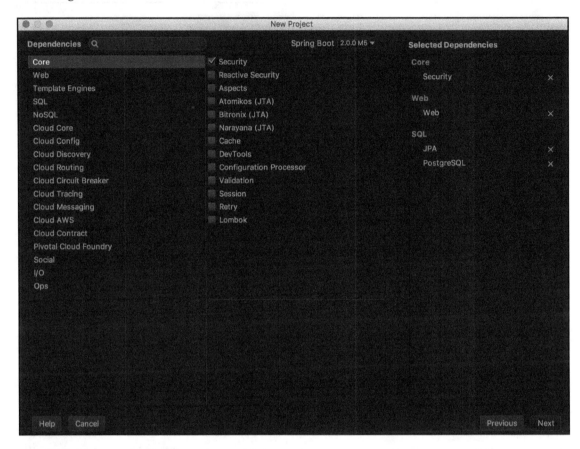

After selecting the appropriate dependencies, click **Next** to continue to the final setup screen. You will be asked to provide a project name and a project location. Fill in `messenger-api` as the project name and select the location where want the project to be saved on your computer. Select **Finish** and wait for the project to be set up. You will be taken to a new IDE window containing the initial project files.

Getting familiar with Spring Boot

Let's take a look at the structure of the initial program files for this Spring Boot application. The following is the screenshot of the structure of your Spring Boot application files:

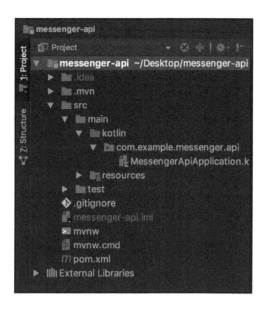

All source files are contained in the `src` directory. This directory contains the core application program files as well as the test programs that are written for the application. Core application program files should be put in the `src/main` directory and test programs are located in `src/test`. The main directory contains two subdirectories. These are the `kotlin` directory and the `resources` directory. All packages and main source files will be placed in this directory over the course of this chapter. More specifically, our program files and packages will be placed within the `com.example.messenger.api` package. Let's have a quick look at the `MessengerApiAplication.kt` file:

The `MessengerApplication.kt` file contains the main function. This is the entry point of every Spring Boot application. This function is called when the application starts. Once it is called, the function runs the Spring application by calling the `SpringApplication.run()` function. This function takes two arguments. The first argument is a class reference and the second is the arguments to be passed to the application upon start.

In the same file, there's a `MessengerApiApplication` class. This class is annotated with the `@SpringBootApplication` annotation. The use of this annotation is equivalent to the combined use of the `@Configuration`, `@EnableAutoConfiguration`, and `@ComponentScan` annotations. Classes annotated with `@Configuration` are sources of bean definitions.

 A bean is an object that is instantiated and assembled by a Spring IoC container.

The `@EnableAutoConfiguration` attribute tells the Spring Boot that you want your Spring application to be automatically configured based on the jar dependencies that you have provided. The `@ComponentScan` annotation configures component scanning directories for use with the `@Configuration` classes.

Over the course of developing Spring Boot applications, it will be necessary to use several annotations for varying reasons. Using these annotations may be overwhelming at first, but with time they will become second nature.

Besides `MessengerApplication.kt`, another important file is the `application.properties` file, located in `src/main/resources`. This file is used for configuring Spring Boot applications properties. Upon opening this file, you will discover that it has no content. This is because we have not yet defined any application configurations or properties. Let's go ahead and add a couple of configurations. Input the following into the `application.properties` file:

```
spring.jpa.generate-ddl=true
spring.jpa.hibernate.ddl-auto=create-drop
```

The `spring.jpa.generate-ddl` property specifies whether the database schema should be generated upon startup of the application. When this property is set to `true`, the schema is generated on application startup, otherwise the schema is not generated. The `spring.jpa.hibernate.ddl-auto` property is used to specify the DDL mode. We use `create-drop` because we want the schema to be created upon application startup and destroyed upon the termination of the application.

We have utilized properties to define the schema of our database but have yet to create an actual database for the `messenger-api`. If you installed pgAdmin along with PostgreSQL, you can easily create a database with the software. If you didn't install pgAdmin, fear not, we can still easily create a database for our application by using PostgreSQL's `createdb` command. Navigate to your terminal and enter the following command:

```
createdb -h localhost --username=<username> --password messenger-api
```

The `-h` flag is used to specify the host name of the machine on which the database server is running. The `--username` flag specifies the username to connect to the server with. The `--password` flag forces a prompt for the specification of a password. The `messenger-api` is the name we are giving to the database being created. Substitute `<username>` with your server username. After you've input the command, click the enter key to run the command. Input a desired password when prompted to do so. A database named `messenger-api` will be created in PostgreSQL.

Now that we have out database set up, we need to connect the Spring Boot application to the database. We can do this with the use of the `spring.datasource.url`, `spring.datasource.username`, and `spring.datasource.password` properties. Add the following configurations to the `application.properties` file:

```
spring.jpa.generate-ddl=true
spring.jpa.hibernate.ddl-auto=create-drop
spring.datasource.url=jdbc:postgresql://localhost:5432/messenger-api
spring.datasource.username=<username>
spring.datasource.password=<password>
```

The `spring.datasource.url` property specifies the JDBC URL via which Spring Boot will connect to the database.
The `spring.datasource.username` and `spring.datasource.password` are the properties used to specify the server username and the password correlating with the specified username. Replace `<username>` and `<password>` with your username and password.

Once you have these properties set up, you are ready to start the Spring Boot application.

Once you have these properties set up, you are ready to start the Spring Boot application. You can run the `messenger-api` application by clicking on the Kotlin logo next to the main function in `MessengerApiApplication.kt` and selecting the **Run** option, as shown in the following screenshot:

Wait for a moment for the project to build. Once the project build process is complete, the application will be started on a Tomcat server.

Let's continue exploring our project files. Locate a `pom.xml` file in the root directory of the project. **POM** stands for **Project Object Model**. On the Apache Maven site, the following is said about the POM: *A Project Object Model or POM is the fundamental unit of work in Maven. It is an XML file that contains information about the project and configuration details used by Maven to build the project.* Once you have located this file, open it. Straightforward, right? That's all fine and dandy, but, for the sake of clarity, here's a brief description of Maven.

Maven

Apache Maven is a software project management and comprehension tool that is based on the concept of the POM. Maven can be used for several purposes, such as project build management and documentation.

Having understood the project files to an extent, we will continue development by implementing some models to cater to the data we identified earlier.

Creating models

Here, we are going to model the data we have identified into suitable entity classes that can be introspected by Spring Boot to build a suitable database schema. The first model we will concern ourselves with is the user model. Create a package named `models` under the `com.example.messenger.api` package. Create a `User.kt` file within the package and input the following code:

```
package com.example.messenger.api.models

import org.hibernate.validator.constraints.Length
import org.springframework.format.annotation.DateTimeFormat
import java.time.Instant
import java.util.*
import javax.persistence.*
import javax.validation.constraints.Pattern
import javax.validation.constraints.Size

@Entity
@Table(name = "`user`")
@EntityListeners(UserListener::class)
class User(
  @Column(unique = true)
  @Size(min = 2)
  var username: String = "",
  @Size(min = 11)
  @Pattern(regexp="^\\(?(\\d{3})\\)?[- ]?(\\d{3})[- ]?(\\d{4})$")
  var phoneNumber: String = "",
  @Size(min = 60, max = 60)
  var password: String = "",
  var status: String = "",
  @Pattern(regexp = "\\A(activated|deactivated)\\z")
  var accountStatus: String = "activated"
)
```

We made use of a lot of annotations in the preceding code block. We will take a look at what each of them does in the order in which they appear. First up, we have the `@Entity` annotation, which indicates that the class is a **Java Persistence API (JPA)** entity. The use of the `@Table` annotation specifies a table name for the entity being represented by the class. This is useful during schema generation. In the case that an `@Table` annotation is not used, the name of the table generated will be the class name. A database table will be created in PostgreSQL with the name `user`. `@EntityListener`, as the name implies, specifies an entity listener for the entity class. We have not yet created a `UserListener` class, but don't worry, we will do that in a little bit.

Now let's take a look at the properties of the `User` class. We added seven class properties in total. The first five are `username`, `password`, `phoneNumber`, `accountStatus`, and `status`; each property represents a type of data we need for a user, as we earlier identified in the *Identifying data* section of this chapter. We have now created our user entity and are ready to proceed. But wait, there's a problem. We need a way to uniquely identify each user that is created. In addition, it is important for future reference to keep track of when new users are added to the messenger platform. After careful consideration, we realize that it is important to have `id` and `createdAt` properties in our entity. You may be wondering—why are we adding `id` and `createdAt` properties to the user entity? After all, we did not specify we needed them earlier on. This is true. But, as we are developing this backend incrementally, we are allowed to make changes and additions when the need arises. Let's go ahead and add these two properties:

```
@Entity
@Table(name = "`user`")
@EntityListeners(UserListener::class)
class User(
    @Column(unique = true)
    @Size(min = 2)
    var username: String = "",
    @Size(min = 8, max = 15)
    @Column(unique = true)
    @Pattern(regexp = "^\\(?(\\d{3})\\)?[- ]?(\\d{3})[- ]?(\\d{4})$")
    var phoneNumber: String = "",
    @Size(min = 60, max = 60)
    var password: String = "",
    var status: String = "available",
    @Pattern(regexp = "\\A(activated|deactivated)\\z")
    var accountStatus: String = "activated",
    @Id
    @GeneratedValue(strategy = GenerationType.AUTO)
    var id: Long = 0,
    @DateTimeFormat
    var createdAt: Date = Date.from(Instant.now())
)
```

Perfect. Now we need to understand what each annotation does. `@Column` is used to specify a property representing a table column. In practice, all properties of entities represent a column in the table. We make use of `@Column(unique = true)` in our code specifically to place a uniqueness constraint on properties. This is useful when we do not want more than one record to have a particular attribute value. `@Size`, as you might have guessed, is used to specify the size of an attribute present in a table. `@Pattern` specifies a pattern that a table attribute must match for it to be valid.

@Id specifies a property that uniquely identifies the entity (the id property, in this case). @GeneratedValue(strategy = GenerationType.AUTO) specifies that we want the id value to be generated automatically. @DateTimeFormat places a timestamp constraint on values to be stored in the created_at column of the user table.

It is time to create a UserListener class. Create a new package named listeners. Add the following UserListener class to the package:

```
package com.example.messenger.api.listeners

import com.example.messenger.api.models.User
import org.springframework.security.crypto.bcrypt.BCryptPasswordEncoder
import javax.persistence.PrePersist
import javax.persistence.PreUpdate

class UserListener {

  @PrePersist
  @PreUpdate
  fun hashPassword(user: User) {
    user.password = BCryptPasswordEncoder().encode(user.password)
  }
}
```

User passwords should never be saved as plain text in a database. For security reasons, they must always be appropriately hashed before being stored. The hashPassword() function performs this hashing procedure by replacing the string value held by the password property of a user object with its hashed equivalent using BCrypt. @PrePersist and @PreUpdate specifies that this function should be called before the persistence or update of a user record in the database.

Now let's create an entity for messages. Go ahead and add a Message class in the models package and add the following code to the class:

```
package com.example.messenger.api.models

import org.springframework.format.annotation.DateTimeFormat
import java.time.Instant
import java.util.*
import javax.persistence.*

@Entity
class Message(
  @ManyToOne(optional = false)
  @JoinColumn(name = "user_id", referencedColumnName = "id")
```

```
    var sender: User? = null,
    @ManyToOne(optional = false)
    @JoinColumn(name = "recipient_id", referencedColumnName = "id")
    var recipient: User? = null,
    var body: String? = "",
    @ManyToOne(optional = false)
    @JoinColumn(name="conversation_id", referencedColumnName = "id")
    var conversation: Conversation? = null,
    @Id @GeneratedValue(strategy = GenerationType.AUTO) var id: Long = 0,
    @DateTimeFormat
    var createdAt: Date = Date.from(Instant.now())
)
```

We made use of some familiar annotations as well as two new ones. As we discussed earlier, every message has a sender as well as a recipient. Both message senders and message recipients are users on the messenger platform, hence the message entity has both sender and recipient properties of the User type. A user can be a sender of many messages as well as a recipient of many messages. These are relationships that need to be implemented. We make use of the @ManyToOne annotation to do this. The many-to-one relationships are not optional, thus we use @ManyToOne(optional = false). @JoinColumn specifies a column for joining an entity association or element collection:

```
@JoinColumn(name = "user_id", referencedColumnName = "id")
var sender: User? = null
```

The code snippet adds a user_id attribute that references the id of a user to the message table.

Upon close inspection, you will notice that a conversation property was used in the Message class. This is because messages sent between users happen in conversation threads. Simply put, every message belongs to a thread. We need to add a Conversation class to our models package, representing the conversation entity:

```
package com.example.messenger.api.models

import org.springframework.format.annotation.DateTimeFormat
import java.time.Instant
import java.util.*
import javax.persistence.*

@Entity
class Conversation(
    @ManyToOne(optional = false)
    @JoinColumn(name = "sender_id", referencedColumnName = "id")
    var sender: User? = null,
    @ManyToOne(optional = false)
```

```
@JoinColumn(name = "recipient_id", referencedColumnName = "id")
var recipient: User? = null,
@Id
@GeneratedValue(strategy = GenerationType.AUTO)
var id: Long = 0,
@DateTimeFormat
val createdAt: Date = Date.from(Instant.now())
) {

    @OneToMany(mappedBy = "conversation", targetEntity = Message::class)
    private var messages: Collection<Message>? = null
}
```

Numerous messages belong to a conversation, so we have a `messages` collection in the body of the `Conversation` class.

We are almost done with the creation of entity models. The only thing left to do is to add appropriate collections for a user's sent and received messages:

```
package com.example.messenger.api.models

import com.example.messenger.api.listeners.UserListener
import org.springframework.format.annotation.DateTimeFormat
import java.time.Instant
import java.util.*
import javax.persistence.*
import javax.validation.constraints.Pattern
import javax.validation.constraints.Size

@Entity
@Table(name = "`user`")
@EntityListeners(UserListener::class)
class User(
  @Column(unique = true)
  @Size(min = 2)
  var username: String = "",
  @Size(min = 8, max = 15)
  @Column(unique = true)
  @Pattern(regexp = "^\\(?(\\d{3})\\)?[- ]?(\\d{3})[- ]?(\\d{4})$")
  var phoneNumber: String = "",
  @Size(min = 60, max = 60)
  var password: String = "",
  var status: String = "available",
  @Pattern(regexp = "\\A(activated|deactivated)\\z")
  var accountStatus: String = "activated",
  @Id
  @GeneratedValue(strategy = GenerationType.AUTO)
```

```
    var id: Long = 0,
    @DateTimeFormat
    var createdAt: Date = Date.from(Instant.now())
) {
    //collection of sent messages
    @OneToMany(mappedBy = "sender", targetEntity = Message::class)
    private var sentMessages: Collection<Message>? = null

    //collection of received messages
    @OneToMany(mappedBy = "recipient", targetEntity = Message::class)
    private var receivedMessages: Collection<Message>? = null
}
```

That's it! We are done creating entities. To help you to understand the entities we have made as well as their relationships, here's an entity relationship diagram (E-R diagram). It shows the entities that we have made and their relationships:

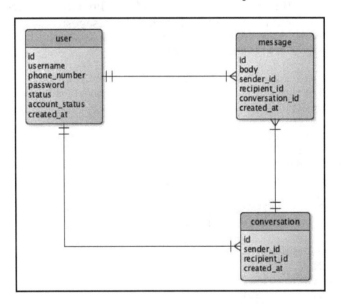

According to the E-R diagram, a user has many messages, a message belongs to a user, a message belongs to a conversation, and a conversation has many messages. In addition, a user has many conversations.

Having created the necessary models, there's only one problem. We have no way to access the data stored by these entities. We need to create repositories to do this.

Creating repositories

Spring Data JPA has is the ability to create repository implementations automatically, at runtime, from a repository interface. We will see how this works by creating a repository to access the User entities. Create a repositories package and include UserRepository.kt in it:

```
package com.example.messenger.api.repositories

import com.example.messenger.api.models.User
import org.springframework.data.repository.CrudRepository

interface UserRepository : CrudRepository<User, Long> {

  fun findByUsername(username: String): User?

  fun findByPhoneNumber(phoneNumber: String): User?
}
```

UserRepository extends the CrudRepository interface. The entity type and id type it works with are specified in the generic parameters of CrudRepository. By extending CrudRepository, UserRepository inherits methods for working with User persistence, such as methods for saving, finding, and deleting User entities.

In addition, Spring JPA allows the declaration of other query functions by the use of method signatures. We exploited this capability to create findByUsername() and findByPhoneNumber() functions.

As we currently have three entities, we need to have three repositories to query them. Create a MessageRepository interface in repositories:

```
package com.example.messenger.api.repositories

import com.example.messenger.api.models.Message
import org.springframework.data.repository.CrudRepository

interface MessageRepository : CrudRepository<Message, Long> {
  fun findByConversationId(conversationId: Long): List<Message>
}
```

Notice the preceding method signature specifies `List<Message>` as its return type. Spring JPA automatically recognizes this and returns a list of `Message` elements when `findByConversationId()` is called.

Lastly, implement a `ConversationRepository` interface:

```
package com.example.messenger.api.repositories

import com.example.messenger.api.models.Conversation
import org.springframework.data.repository.CrudRepository

interface ConversationRepository : CrudRepository<Conversation, Long> {
    fun findBySenderId(id: Long): List<Conversation>

    fun findByRecipientId(id: Long): List<Conversation>

    fun findBySenderIdAndRecipientId(senderId: Long,
        recipientId: Long): Conversation?
}
```

As we now have our entities set and the necessary repositories to query these entities, we can start work on implementing the business logic of the messenger backend. This will require us to learn about services and service implementations.

Services and service implementations

A service implementation is a spring bean that is annotated by `@Service`. Business logic for spring applications is most commonly put in a service implementation. A *service*, on the other hand, is an interface with function signatures for application behavior that must be implemented by implementing classes. A simple way to recall the differentiation between the two is to keep in mind that a service is an interface and a service implementation is a class that implements a service.

Now to create some services and service implementations. Create a `service` package. We will add both services and service implementations in here. Create a `UserService` interface as with the following codes within the package:

```
package com.example.messenger.api.services

import com.example.messenger.api.models.User

interface UserService {
    fun attemptRegistration(userDetails: User): User
```

```
fun listUsers(currentUser): List<User>

fun retrieveUserData(username: String): User?

fun retrieveUserData(id: Long): User?

fun usernameExists(username: String): Boolean
}
```

In the preceding `UserService` interface, we have defined functions that must be declared by classes that implement `UserService`. That's it! `UserService` is ready to be implemented. Now to create an implementation of the service. Add a `UserServiceImpl` class to the services package. We are going to implement `UserService` and as such we need overriding functions for `attemptRegistration()`, `listUsers()`, `retrieveUserData()`, and `usernameExists()`:

```
package com.example.messenger.api.services

import com.example.messenger.api.exceptions.InvalidUserIdException
import com.example.messenger.api.exceptions.UserStatusEmptyException
import com.example.messenger.api.exceptions.UsernameUnavailableException
import com.example.messenger.api.models.User
import com.example.messenger.api.repositories.UserRepository
import org.springframework.stereotype.Service


@Service
class UserServiceImpl(val repository: UserRepository) : UserService {
  @Throws(UsernameUnavailableException::class)
  override fun attemptRegistration(userDetails: User): User {
    if (!usernameExists(userDetails.username)) {
      val user = User()
      user.username = userDetails.username
      user.phoneNumber = userDetails.phoneNumber
      user.password = userDetails.password
      repository.save(user)
      obscurePassword(user)
      return user
    }
    throw UsernameUnavailableException("The username
                ${userDetails.username} is unavailable.")
  }

  @Throws(UserStatusEmptyException::class)
  fun updateUserStatus(currentUser: User, updateDetails: User): User {
    if (!updateDetails.status.isEmpty()) {
```

```kotlin
            currentUser.status = updateDetails.status
            repository.save(currentUser)
            return  currentUser
        }
        throw UserStatusEmptyException()
    }

    override fun listUsers(currentUser: User): List<User> {
        return repository.findAll().mapTo(ArrayList(), { it })
                        .filter{ it != currentUser }
    }

    override fun retrieveUserData(username: String): User? {
        val user = repository.findByUsername(username)
        obscurePassword(user)
        return user
    }

    @Throws(InvalidUserIdException::class)
    override fun retrieveUserData(id: Long): User {
        val userOptional = repository.findById(id)
        if (userOptional.isPresent) {
            val user = userOptional.get()
            obscurePassword(user)
            return user
        }
        throw InvalidUserIdException("A user with an id of '$id'
                                    does not exist.")
    }

    override fun usernameExists(username: String): Boolean {
        return repository.findByUsername(username) != null
    }

    private fun obscurePassword(user: User?) {
        user?.password = "XXX XXXX XXX"
    }
}
```

In the primary constructor definition of UserServiceImpl, an instance of
UserRepository was specified as a required argument. You don't need to worry about
passing such an argument yourself. Spring recognizes that UserServiceImpl needs a
UserRepository instance and provides the class with one via dependency injection. In
addition to the functions implemented, we declared an obscurePassword() function that
simply hashed passwords within a User entity with XXX XXXX XXX.

Still in the spirit of service and service implementation creation, let's go ahead and add some for messages and conversations. Add a `MessageService` interface to the `service`:

```
package com.example.messenger.api.services

import com.example.messenger.api.models.Message
import com.example.messenger.api.models.User

interface MessageService {

  fun sendMessage(sender: User, recipientId: Long,
                  messageText: String): Message
}
```

We added a single method signature for `sendMessage()` that must be overriden by `MessageServiceImpl`. The following is the message service implementation:

```
package com.example.messenger.api.services

import com.example.messenger.api.exceptions.MessageEmptyException
import
com.example.messenger.api.exceptions.MessageRecipientInvalidException
import com.example.messenger.api.models.Conversation
import com.example.messenger.api.models.Message
import com.example.messenger.api.models.User
import com.example.messenger.api.repositories.ConversationRepository
import com.example.messenger.api.repositories.MessageRepository
import com.example.messenger.api.repositories.UserRepository
import org.springframework.stereotype.Service

@Service
class MessageServiceImpl(val repository: MessageRepository,
            val conversationRepository: ConversationRepository,
            val conversationService: ConversationService,
            val userRepository: UserRepository) : MessageService {

  @Throws(MessageEmptyException::class,
          MessageRecipientInvalidException::class)
  override fun sendMessage(sender: User, recipientId: Long,
                           messageText: String): Message {
    val optional = userRepository.findById(recipientId)

      if (optional.isPresent) {
        val recipient = optional.get()

        if (!messageText.isEmpty()) {
          val conversation: Conversation = if (conversationService
```

```
                       .conversationExists(sender, recipient)) {
          conversationService.getConversation(sender, recipient)
                            as Conversation
      } else {
        conversationService.createConversation(sender, recipient)
      }
      conversationRepository.save(conversation)

      val message = Message(sender, recipient, messageText,
                            conversation)
      repository.save(message)
      return message
    }
  } else {
    throw MessageRecipientInvalidException("The recipient id
                        '$recipientId' is invalid.")
  }
  throw MessageEmptyException()
  }
}
```

The preceding implementation of `sendMessage()` first checks whether the message content is empty. If not, then the function checks whether there exists an active conversation between the sender and the recipient. If there is one, it is retrieved and stored in `conversation`, otherwise a new `Conversation` is created between the two users and stored in `conversation`. The conversation is then saved and the message is created and saved.

`ConversationService` and `ConversationServiceImpl` can now be implemented. Create a `ConversationService` interface in `services` and add the following code:

```
package com.example.messenger.api.services

import com.example.messenger.api.models.Conversation
import com.example.messenger.api.models.User

interface ConversationService {

    fun createConversation(userA: User, userB: User): Conversation
    fun conversationExists(userA: User, userB: User): Boolean
    fun getConversation(userA: User, userB: User): Conversation?
    fun retrieveThread(conversationId: Long): Conversation
    fun listUserConversations(userId: Long): List<Conversation>
    fun nameSecondParty(conversation: Conversation, userId: Long): String
}
```

We have added six function signatures for now. They are createConversation(), conversationExists(), getConversation(), retrieveThread(), listUserConversations(), and nameSecondParty(). Now we shall add ConversationServiceImpl to services and implement the first three methods createConversation(), conversationExists() and getConversation(). This implementation is shown in the following code snippet:

```
package com.example.messenger.api.services

import com.example.messenger.api.exceptions.ConversationIdInvalidException
import com.example.messenger.api.models.Conversation
import com.example.messenger.api.models.User
import com.example.messenger.api.repositories.ConversationRepository
import org.springframework.stereotype.Service

@Service
class ConversationServiceImpl(val repository: ConversationRepository) :
        ConversationService {

  override fun createConversation(userA: User, userB: User):
                Conversation {
    val conversation = Conversation(userA, userB)
    repository.save(conversation)
    return  conversation
  }

  override fun conversationExists(userA: User, userB: User): Boolean {
    return if (repository.findBySenderIdAndRecipientId
            (userA.id, userB.id) != null)
    true
    else repository.findBySenderIdAndRecipientId
            (userB.id, userA.id) != null
  }

  override fun getConversation(userA: User, userB: User): Conversation? {
    return when {
      repository.findBySenderIdAndRecipientId(userA.id,
                        userB.id) != null ->
      repository.findBySenderIdAndRecipientId(userA.id, userB.id)
      repository.findBySenderIdAndRecipientId(userB.id,
                        userA.id) != null ->
      repository.findBySenderIdAndRecipientId(userB.id, userA.id)
      else -> null
    }
  }

}
```

```
}
```

Having added the first three methods, go ahead and include the remaining three methods, `retrieveThread()`, `listUserConversations()`, and `nameSecondParty()`, below to `ConversationServiceImpl`:

```
override fun retrieveThread(conversationId: Long): Conversation {
  val conversation = repository.findById(conversationId)

  if (conversation.isPresent) {
    return conversation.get()
  }
  throw ConversationIdInvalidException("Invalid conversation id
                                      '$conversationId'")
}

override fun listUserConversations(userId: Long):
              ArrayList<Conversation> {
  val conversationList: ArrayList<Conversation> = ArrayList()
  conversationList.addAll(repository.findBySenderId(userId))
  conversationList.addAll(repository.findByRecipientId(userId))

  return conversationList
}

override fun nameSecondParty(conversation: Conversation,
                            userId: Long): String {
  return if (conversation.sender?.id == userId) {
    conversation.recipient?.username as String
  } else {
    conversation.sender?.username as String
  }
}
}
```

You might have noticed that we threw exceptions of different types several times within service implementation classes. As we have not yet created these exceptions, we will need to do so. In addition, we need to create an `ExceptionHandler` for each of these exceptions. These exception handlers will send appropriate error responses to clients in scenarios in which exceptions are thrown.

Create an `exceptions` package and add an `AppExceptions.kt` file to it. Include the following code into the file:

```
package com.example.messenger.api.exceptions

class UsernameUnavailableException(override val message: String) :
RuntimeException()

class InvalidUserIdException(override val message: String) :
RuntimeException()

class MessageEmptyException(override val message: String = "A message
cannot be empty.") : RuntimeException()

class MessageRecipientInvalidException(override val message: String) :
RuntimeException()

class ConversationIdInvalidException(override val message: String) :
RuntimeException()

class UserDeactivatedException(override val message: String) :
RuntimeException()

class UserStatusEmptyException(override val message: String = "A user's
status cannot be empty") : RuntimeException()
```

Each exception extends `RuntimeException` as they occur during the server runtime. All exceptions also possess a `message` property. As the name implies, this is the exception message. Now that our exceptions have been added, we need to create controller advice classes. `ControllerAdvice` classes are used to handle errors that occur within a Spring application. They are created using the `@ControllerAdvice` annotation. In addition, a controller advice is a type of Spring component. Let's create a controller advice class to handle some of the preceding exceptions.

Looking at `UsernameUnavailableException`, `InvalidUserIdException`, and `UserStatusEmptyException`, we notice that these three exceptions are all pertaining to a user. As such, let's name the controller advice that caters to all these exceptions `UserControllerAdvice`. Create a `components` package and add a the following `UserControllerAdvice` class to it:

```
package com.example.messenger.api.components

import com.example.messenger.api.constants.ErrorResponse
import com.example.messenger.api.constants.ResponseConstants
import com.example.messenger.api.exceptions.InvalidUserIdException
```

```
import com.example.messenger.api.exceptions.UserStatusEmptyException
import com.example.messenger.api.exceptions.UsernameUnavailableException
import org.springframework.http.ResponseEntity
import org.springframework.web.bind.annotation.ControllerAdvice
import org.springframework.web.bind.annotation.ExceptionHandler

@ControllerAdvice
class UserControllerAdvice {

  @ExceptionHandler(UsernameUnavailableException::class)
  fun usernameUnavailable(usernameUnavailableException:
                          UsernameUnavailableException):
    ResponseEntity<ErrorResponse> {
    val res = ErrorResponse(ResponseConstants.USERNAME_UNAVAILABLE
                    .value, usernameUnavailableException.message)
    return ResponseEntity.unprocessableEntity().body(res)
  }

  @ExceptionHandler(InvalidUserIdException::class)
  fun invalidId(invalidUserIdException: InvalidUserIdException):
    ResponseEntity<ErrorResponse> {
    val res = ErrorResponse(ResponseConstants.INVALID_USER_ID.value,
                        invalidUserIdException.message)
    return ResponseEntity.badRequest().body(res)
  }

  @ExceptionHandler(UserStatusEmptyException::class)
  fun statusEmpty(userStatusEmptyException: UserStatusEmptyException):
    ResponseEntity<ErrorResponse> {
    val res = ErrorResponse(ResponseConstants.EMPTY_STATUS.value,
                        userStatusEmptyException.message)
    return ResponseEntity.unprocessableEntity().body(res)
  }
}
```

We've just defined functions to cater to each of the three exceptions that can occur and annotated each of the functions with an `@ExceptionHanlder()` annotation. `@ExceptionHanlder()` takes a class reference to the exception that is being handled by the function. Each function takes a single argument that is an instance of the exception thrown. In addition, all the defined functions return a `ResponseEntity<ErrorResponse>` instance. A response entity represents the entire HTTP response sent to the client.

`ErrorResponse` has not yet been created. Create a `constants` package and add the following `ErrorResponse` class to it:

```
package com.example.messenger.api.constants

class ErrorResponse(val errorCode: String, val errorMessage: String)
```

`ErrorResponse` is a simple class with two properties: `errorCode` and `errorMessage`. Before we continue, go ahead and add the following `ResponseConstants` enum class to the `constants` package:

```
package com.example.messenger.api.constants

enum class ResponseConstants(val value: String) {
  SUCCESS("success"), ERROR("error"),
  USERNAME_UNAVAILABLE("USR_0001"),
  INVALID_USER_ID("USR_002"),
  EMPTY_STATUS("USR_003"),
  MESSAGE_EMPTY("MES_001"),
  MESSAGE_RECIPIENT_INVALID("MES_002"),
  ACCOUNT_DEACTIVATED("GLO_001")
}
```

Now, let's create three more controller advice classes. These classes are `MessageControllerAdvice`, `ConversationControllerAdvice`, and `RestControllerAdvice`. `RestControllerAdvice` will define exception handlers for errors that can happen anywhere within the server over the course of runtime.

The following is the `MessageControllerAdvice` class:

```
package com.example.messenger.api.components

import com.example.messenger.api.constants.ErrorResponse
import com.example.messenger.api.constants.ResponseConstants
import com.example.messenger.api.exceptions.MessageEmptyException
import
com.example.messenger.api.exceptions.MessageRecipientInvalidException
import org.springframework.http.ResponseEntity
import org.springframework.web.bind.annotation.ControllerAdvice
import org.springframework.web.bind.annotation.ExceptionHandler

@ControllerAdvice
class MessageControllerAdvice {
  @ExceptionHandler(MessageEmptyException::class)
  fun messageEmpty(messageEmptyException: MessageEmptyException):
    ResponseEntity<ErrorResponse> {
```

```kotlin
    //ErrorResponse object creation
    val res = ErrorResponse(ResponseConstants.MESSAGE_EMPTY.value,
                            messageEmptyException.message)

    // Returning ResponseEntity containing appropriate ErrorResponse
    return ResponseEntity.unprocessableEntity().body(res)
  }

  @ExceptionHandler(MessageRecipientInvalidException::class)
    fun messageRecipientInvalid(messageRecipientInvalidException:
                                MessageRecipientInvalidException):
        ResponseEntity<ErrorResponse> {
    val res = ErrorResponse(ResponseConstants.MESSAGE_RECIPIENT_INVALID
                  .value, messageRecipientInvalidException.message)
    return ResponseEntity.unprocessableEntity().body(res)
  }
}
```

Next, add the `ConversationControllerAdvice` class, which is as follows:

```kotlin
package com.example.messenger.api.components

import com.example.messenger.api.constants.ErrorResponse
import com.example.messenger.api.exceptions.ConversationIdInvalidException
import org.springframework.http.ResponseEntity
import org.springframework.web.bind.annotation.ControllerAdvice
import org.springframework.web.bind.annotation.ExceptionHandler

@ControllerAdvice
class ConversationControllerAdvice {
  @ExceptionHandler
  fun conversationIdInvalidException(conversationIdInvalidException:
          ConversationIdInvalidException): ResponseEntity<ErrorResponse> {
    val res = ErrorResponse("", conversationIdInvalidException.message)
    return ResponseEntity.unprocessableEntity().body(res)
  }
}
```

Finally, add the `RestControllerAdvice` class:

```kotlin
package com.example.messenger.api.components

import com.example.messenger.api.constants.ErrorResponse
import com.example.messenger.api.constants.ResponseConstants
import com.example.messenger.api.exceptions.UserDeactivatedException
import org.springframework.http.HttpStatus
import org.springframework.http.ResponseEntity
import org.springframework.web.bind.annotation.ControllerAdvice
```

```
import org.springframework.web.bind.annotation.ExceptionHandler

@ControllerAdvice
class RestControllerAdvice {

  @ExceptionHandler(UserDeactivatedException::class)
  fun userDeactivated(userDeactivatedException:
                      UserDeactivatedException):
    ResponseEntity<ErrorResponse> {
    val res = ErrorResponse(ResponseConstants.ACCOUNT_DEACTIVATED
                  .value, userDeactivatedException.message)

    // Return an HTTP 403 unauthorized error response
    return ResponseEntity(res, HttpStatus.UNAUTHORIZED)
  }
}
```

We have implemented our business logic and we are almost ready to facilitate HTTP request entries via REST endpoints into our API. Before we do that, we must secure our API.

Restricting API access

From a security standpoint, it is a huge taboo to permit just anyone to have access to RESTful API resources. We must devise a way to restrict access to our server to only registered and logged-in users. We will do this using **Spring Security** and **JSON Web Tokens (JWTs)**.

Spring Security

Spring Security is a highly customizable access-control framework for Spring applications. It is the accepted standard for securing applications built with Spring. As it is selected to add the Security dependency at the start of this project's creation, we do not need to add the Spring Security dependency to `pom.xml` as it has already been added.

JSON Web Tokens

According to the JWT website (`https://tools.ietf.org/html/rfc7519`), *JSON Web Tokens are an open, industry standard method for representing claims securely between two parties*. JWT allows you to decode, verify, and generate JWT. JWTs can be used easily with Spring Boot to implement authentication in applications. The following sections will demonstrate how to use a JWT and Spring Security combination to secure the messenger backend.

The first thing that must be done in order for you to get started with JWTs in a Spring application is add its dependency to the project `pom.xml` file:

```
<dependencies>
  ...
  <dependency>
    <groupId>io.jsonwebtoken</groupId>
    <artifactId>jjwt</artifactId>
    <version>0.7.0</version>
  </dependency>
</dependencies>
```

Upon including a new maven dependency in `pom.xml`, IntelliJ will ask you to import the new dependencies:

Click on **Import Changes** once the prompt shows up, and the JWT dependency will be imported into the project.

Configuring web security

The first thing we need to do is create a custom web security configuration. Create a config package in com.example.messenger.api. Add a WebSecurityConfig class to the package and input the following code:

```
package com.example.messenger.api.config

import com.example.messenger.api.filters.JWTAuthenticationFilter
import com.example.messenger.api.filters.JWTLoginFilter
import com.example.messenger.api.services.AppUserDetailsService
import org.springframework.context.annotation.Configuration
import org.springframework.http.HttpMethod
import
org.springframework.security.config.annotation.authentication.builders.Auth
enticationManagerBuilder
import
org.springframework.security.config.annotation.web.builders.HttpSecurity
import
org.springframework.security.config.annotation.web.configuration.EnableWebS
ecurity
import
org.springframework.security.config.annotation.web.configuration.WebSecurit
yConfigurerAdapter
import org.springframework.security.core.userdetails.UserDetailsService
import org.springframework.security.crypto.bcrypt.BCryptPasswordEncoder
import
org.springframework.security.web.authentication.UsernamePasswordAuthenticat
ionFilter

@Configuration
@EnableWebSecurity
class WebSecurityConfig(val userDetailsService: AppUserDetailsService)
        : WebSecurityConfigurerAdapter() {

  @Throws(Exception::class)
  override fun configure(http: HttpSecurity) {
    http.csrf().disable().authorizeRequests()
        .antMatchers(HttpMethod.POST, "/users/registrations")
        .permitAll()
        .antMatchers(HttpMethod.POST, "/login").permitAll()
        .anyRequest().authenticated()
        .and()
```

Let's `Filter` the /login requests:

```
.addFilterBefore(JWTLoginFilter("/login",
        authenticationManager()),
    UsernamePasswordAuthenticationFilter::class.java)
```

Let's filter other requests to check the presence of JWT in header:

```
.addFilterBefore(JWTAuthenticationFilter(),
        UsernamePasswordAuthenticationFilter::class.java)
}

@Throws(Exception::class)
override fun configure(auth: AuthenticationManagerBuilder) {
    auth.userDetailsService<UserDetailsService>(userDetailsService)
        .passwordEncoder(BCryptPasswordEncoder())
}
}
```

`WebSecurityConfig` is annotated with `@EnableWebSecurity`. This enables Spring Security's web security support. In addition, `WebSecurityConfig` extends `WebSecurityConfigurerAdapter` and overrides some of its `configure()` methods to add some customization to the web security config.

The `configure(HttpSecurity)` method configures which URL paths are to be secured and which shouldn't be. In `WebSecurityConfig`, we permitted all `POST` requests to the `/users/registrations` and `/login` paths. These two endpoints don't need to be secured, as a user cannot be authenticated prior to login or his registration on the platform. In addition, we added filters for requests. Requests to `/login` will be filtered by `JWTLoginFilter` (we have yet to implement this); all requests that are unauthenticated and unpermitted will be filtered by `JWTAuthenticationFilter` (we have yet to implement this, too).

`configure(AuthenticationManagerBuilder)` sets up the `UserDetailsService` and specifies a password encoder to be used.

There are a number of classes that we made use of that we have not implemented yet. We will start by implementing `JWTLoginFilter`. Create a new package named `filters` and add the following `JWTLoginFilter` class:

```
package com.example.messenger.api.filters

import com.example.messenger.api.security.AccountCredentials
import com.example.messenger.api.services.TokenAuthenticationService
import com.fasterxml.jackson.databind.ObjectMapper
```

```
import org.springframework.security.authentication.AuthenticationManager
import
org.springframework.security.authentication.UsernamePasswordAuthenticationT
oken
import org.springframework.security.core.Authentication
import org.springframework.security.core.AuthenticationException
import
org.springframework.security.web.authentication.AbstractAuthenticationProce
ssingFilter
import org.springframework.security.web.util.matcher.AntPathRequestMatcher

import javax.servlet.FilterChain
import javax.servlet.ServletException
import javax.servlet.http.HttpServletRequest
import javax.servlet.http.HttpServletResponse
import java.io.IOException

class JWTLoginFilter(url: String, authManager: AuthenticationManager) :
    AbstractAuthenticationProcessingFilter(AntPathRequestMatcher(url)){

  init {
    authenticationManager = authManager
  }

  @Throws(AuthenticationException::class, IOException::class,
          ServletException::class)
  override fun attemptAuthentication( req: HttpServletRequest,
                      res: HttpServletResponse): Authentication{
    val credentials = ObjectMapper()
        .readValue(req.inputStream, AccountCredentials::class.java)
    return authenticationManager.authenticate(
      UsernamePasswordAuthenticationToken(
        credentials.username,
        credentials.password,
        emptyList()
      )
    )
  }

  @Throws(IOException::class, ServletException::class)
  override fun successfulAuthentication(
              req: HttpServletRequest,
              res: HttpServletResponse, chain: FilterChain,
              auth: Authentication) {
    TokenAuthenticationService.addAuthentication(res, auth.name)
  }
}
```

`JWTLoginFilter` takes a string URL and an `AuthenticationManager` instance as arguments to its primary constructor. You can also see it extends `AbstractAuthenticationProcessingFilter`. This filter intercepts incoming HTTP requests to the server and attempts to authenticate them. `attemptAuthentication()` performs the actual authentication process. It uses an `ObjectMapper()` instance to read the credentials present in the `via` HTTP request, after which `authenticationManager` is used to authenticate the request. `AccountCredentials` is another class that we have yet to implement. Create a new package, called `security`, and add an `AccountCredentials.kt` file to it:

```
package com.example.messenger.api.security


class AccountCredentials {
    lateinit var username: String
    lateinit var password: String
}
```

We have variables for a `username` and `password` because these are what will be used to authenticate the user.

The `SuccessfulAuthentication()` method is called upon successful authentication of a user. The only task done in the function is the addition of authentication tokens to the `Authorization` header of the HTTP response. The actual addition of this header is done by `TokenAuthenticationService.addAuthentication()`. Let's add this service to our `services` package:

```
package com.example.messenger.api.services
import io.jsonwebtoken.Jwts
import io.jsonwebtoken.SignatureAlgorithm
import
org.springframework.security.authentication.UsernamePasswordAuthenticationT
oken
import org.springframework.security.core.Authentication
import org.springframework.security.core.GrantedAuthority

import javax.servlet.http.HttpServletRequest
import javax.servlet.http.HttpServletResponse
import java.util.Date

import java.util.Collections.emptyList

internal object TokenAuthenticationService {
    private val TOKEN_EXPIRY: Long = 864000000
    private val SECRET = "$78gr43g7g8feb8we"
```

```
private val TOKEN_PREFIX = "Bearer"
private val AUTHORIZATION_HEADER_KEY = "Authorization"

fun addAuthentication(res: HttpServletResponse, username: String) {
  val JWT = Jwts.builder()
              .setSubject(username)
              .setExpiration(Date(System.currentTimeMillis() +
                                  TOKEN_EXPIRY))
              .signWith(SignatureAlgorithm.HS512, SECRET)
              .compact()
  res.addHeader(AUTHORIZATION_HEADER_KEY, "$TOKEN_PREFIX $JWT")
}

fun getAuthentication(request: HttpServletRequest): Authentication? {
  val token = request.getHeader(AUTHORIZATION_HEADER_KEY)
  if (token != null) {
```

Let's parse the token:

```
    val user = Jwts.parser().setSigningKey(SECRET)
                .parseClaimsJws(token.replace(TOKEN_PREFIX, ""))
                .body.subject

  if (user != null)
    return UsernamePasswordAuthenticationToken(user, null,
                          emptyList<GrantedAuthority>())
  }
  return null
 }
}
```

As the names imply, `addAuthentication()` adds an authentication token to the `Authorization` header of the HTTP response and `getAuthentication()` authenticates the user.

Now let's add `JWTAuthenticationFilter` to the `filters` package. Add the following `JWTAuthenticationFilter` class to the `filters` package:

```
package com.example.messenger.api.filters

import com.example.messenger.api.services.TokenAuthenticationService
import org.springframework.security.core.context.SecurityContextHolder
import org.springframework.web.filter.GenericFilterBean
import javax.servlet.FilterChain
import javax.servlet.ServletException
import javax.servlet.ServletRequest
import javax.servlet.ServletResponse
```

```kotlin
import javax.servlet.http.HttpServletRequest
import java.io.IOException

class JWTAuthenticationFilter : GenericFilterBean() {

  @Throws(IOException::class, ServletException::class)
  override fun doFilter(request: ServletRequest,
                        response: ServletResponse,
                        filterChain: FilterChain) {
    val authentication = TokenAuthenticationService
            .getAuthentication(request as HttpServletRequest)
    SecurityContextHolder.getContext().authentication = authentication
    filterChain.doFilter(request, response)
  }
}
```

The doFilter() function of the JWTAuthenticationFilter is called by the container each time a request/response pair is passed through the filter chain as a result of a client request for a resource. The FilterChain instance passed in to doFilter() allows the filter to pass the request and response on to the next entity in the filter chain.

Finally, we need to implement the AppUserDetailsService class as usual, we will put this in the services package of the project:

```kotlin
package com.example.messenger.api.services

import com.example.messenger.api.repositories.UserRepository
import org.springframework.security.core.GrantedAuthority
import org.springframework.security.core.authority.SimpleGrantedAuthority
import org.springframework.security.core.userdetails.User
import org.springframework.security.core.userdetails.UserDetails
import org.springframework.security.core.userdetails.UserDetailsService
import
org.springframework.security.core.userdetails.UsernameNotFoundException
import org.springframework.stereotype.Component
import java.util.ArrayList

@Component
class AppUserDetailsService(val userRepository: UserRepository) :
UserDetailsService {

  @Throws(UsernameNotFoundException::class)
  override fun loadUserByUsername(username: String): UserDetails {
    val user = userRepository.findByUsername(username) ?:
                throw UsernameNotFoundException("A user with the
                              username $username doesn't exist")
```

```
    return User(user.username, user.password,
            ArrayList<GrantedAuthority>())
  }
}
```

`loadUsername(String)` attempts to load the `UserDetails` of a user matching the `username` passed to the function. If the user matching the provided username cannot be found, a `UsernameNotFoundException` is thrown.

And, just like that, we have successfully configured Spring Security. We are now ready to expose some API functionality via RESTful endpoints with the use of controllers.

Accessing server resources via RESTful endpoints

So far, we have created models, components, services, and service implementations, as well as integrated Spring Security into the messenger application. One thing we have not done is actually created any means by which external clients can communicate with the messenger API. We are going to do this by creating controller classes that handles requests from different HTTP request paths. As always, the first thing we must do is create a package to contain the controllers we are about to create. Create a `controllers` package now.

The first controller we will implement is the `UserController`. This controller maps HTTP requests pertaining to a user resource to in-class actions that handle and respond to the HTTP request. First and foremost, we need an endpoint to facilitate the registration of new users. We will call the action that handles such a registration request `create`. The following is the `UserController` code with the `create` action:

```
package com.example.messenger.api.controllers

import com.example.messenger.api.models.User
import com.example.messenger.api.repositories.UserRepository
import com.example.messenger.api.services.UserServiceImpl
import org.springframework.http.ResponseEntity
import org.springframework.validation.annotation.Validated
import org.springframework.web.bind.annotation.*
import javax.servlet.http.HttpServletRequest

@RestController
@RequestMapping("/users")
class UserController(val userService: UserServiceImpl,
                     val userRepository: UserRepository) {

  @PostMapping
  @RequestMapping("/registrations")
```

```
fun create(@Validated @RequestBody userDetails: User):
            ResponseEntity<User> {
    val user = userService.attemptRegistration(userDetails)
    return ResponseEntity.ok(user)
  }
}
```

The controller class is annotated with `@RestController` and `@RequestMapping`. The `@RestController` annotation specifies that a class is a REST controller. `@RequestMapping`, as it is used with the `UserController` class earlier, maps all requests with paths starting with `/users` to `UserController`.

The `create` function is annotated with `@PostMapping` and `@RequestMapping("/registrations")`. The combination of these two annotations maps all POST requests with the `/users/registrations` path to the create function. A `User` instance annotated with `@Validated` and `@RequestBody` is passed to `create`. `@RequestBody` binds the JSON values sent in the body of the POST request to `userDetails`. `@Validated` ensures that the JSON parameters are validated. Now that we have an endpoint up and running, let's test it out. Start the application and navigate to your terminal window. Send a request to the messenger API using CURL, as follows:

```
curl -H "Content-Type: application/json" -X POST -d
'{"username":"kevin.stacey",
  "phoneNumber":"5472457893",
  "password":"Hello123"}'
 http://localhost:8080/users/registrations
```

The server will create the user and send you a response similar to the following:

```
{
  "username":"kevin.stacey",
  "phoneNumber":"5472457893",
  "password":"XXX XXXX XXX",
  "status":"available",
  "accountStatus":"activated",
  "id":6,"createdAt":1508579448634
}
```

That's all fine and good, but we can see there are a number of unwanted values in the HTTP response, such as the `password` and `accountStatus` response parameters. In addition to this, we'd like for `createdAt` to contain a human-readable date. We are going to do all these things using an assembler and a value object.

First, let's make the value object. The value object we are creating is going to contain the data of the user that we want to be sent to the client in its appropriate form and nothing more. Create a `helpers.objects` package with a `ValueObjects.kt` file in it:

```
package com.example.messenger.api.helpers.objects

data class UserVO(
    val id: Long,
    val username: String,
    val phoneNumber: String,
    val status: String,
    val createdAt: String
)
```

As you can see, `UserVO` is a data class that models the information we want to be sent to the user and nothing more. While we are at it, let's add value objects for some other responses we will cater for later, to avoid coming back to this file:

```
package com.example.messenger.api.helpers.objects

data class UserVO(
    val id: Long,
    val username: String,
    val phoneNumber: String,
    val status: String,
    val createdAt: String
)

data class UserListVO(
    val users: List<UserVO>
)

data class MessageVO(
    val id: Long,
    val senderId: Long?,
    val recipientId: Long?,
    val conversationId: Long?,
    val body: String?,
    val createdAt: String
)

data class ConversationVO(
    val conversationId: Long,
    val secondPartyUsername: String,
    val messages: ArrayList<MessageVO>
)
```

```
data class ConversationListVO(
    val conversations: List<ConversationVO>
)
```

Now that we have the required value objects set, let's create an assembler for UserVO. An assembler is simply a component that assembles a required object value. We will call the assembler we are creating UserAssembler. As it's a component, it belongs in the components package:

```
package com.example.messenger.api.components

import com.example.messenger.api.helpers.objects.UserListVO
import com.example.messenger.api.helpers.objects.UserVO
import com.example.messenger.api.models.User
import org.springframework.stereotype.Component

@Component
class UserAssembler {

    fun toUserVO(user: User): UserVO {
        return UserVO(user.id, user.username, user.phoneNumber,
                      user.status, user.createdAt.toString())
    }

    fun toUserListVO(users: List<User>): UserListVO {
        val userVOList = users.map { toUserVO(it) }
        return  UserListVO(userVOList)
    }
}
```

The assembler has a single toUserVO() function that takes a User as its argument and returns a corresponding UserVO. toUserListVO() takes a list of User instances and returns a corresponding UserListVO.

Now let's edit the create endpoint to make use of UserAssembler and UserVO:

```
package com.example.messenger.api.controllers

import com.example.messenger.api.components.UserAssembler
import com.example.messenger.api.helpers.objects.UserVO
import com.example.messenger.api.models.User
import com.example.messenger.api.repositories.UserRepository
import com.example.messenger.api.services.UserServiceImpl
import org.springframework.http.ResponseEntity
import org.springframework.validation.annotation.Validated
import org.springframework.web.bind.annotation.*
import javax.servlet.http.HttpServletRequest
```

```
@RestController
@RequestMapping("/users")
class UserController(val userService: UserServiceImpl,
                     val userAssembler: UserAssembler,
                     val userRepository: UserRepository) {

  @PostMapping
  @RequestMapping("/registrations")
  fun create(@Validated @RequestBody userDetails: User):
          ResponseEntity<UserVO> {
    val user = userService.attemptRegistration(userDetails)
    return ResponseEntity.ok(userAssembler.toUserVO(user))
  }
}
```

Restart the server and send a new request to register a User. We will get a response that is much more appropriate from the API:

```
{
  "id":6,
  "username":"kevin.stacey",
  "phoneNumber":"5472457893",
  "status":"available",
  "createdAt":"Sat Oct 21 11:11:36 WAT 2017"
}
```

Let's wrap up our endpoint creation process by creating all the necessary endpoints for the messenger Android application. Firstly, let's add endpoints to show the details of a User, list all users, get the details of the current user, and update the status of a User to UserController:

```
package com.example.messenger.api.controllers

import com.example.messenger.api.components.UserAssembler
import com.example.messenger.api.helpers.objects.UserListVO
import com.example.messenger.api.helpers.objects.UserVO
import com.example.messenger.api.models.User
import com.example.messenger.api.repositories.UserRepository
import com.example.messenger.api.services.UserServiceImpl
import org.springframework.http.ResponseEntity
import org.springframework.validation.annotation.Validated
import org.springframework.web.bind.annotation.*
import javax.servlet.http.HttpServletRequest

@RestController
@RequestMapping("/users")
class UserController(val userService: UserServiceImpl,
```

```kotlin
                    val userAssembler: UserAssembler,
                    val userRepository: UserRepository) {

    @PostMapping
    @RequestMapping("/registrations")
    fun create(@Validated @RequestBody userDetails: User):
                ResponseEntity<UserVO> {
      val user = userService.attemptRegistration(userDetails)
      return ResponseEntity.ok(userAssembler.toUserVO(user))
    }

    @GetMapping
    @RequestMapping("/{user_id}")
    fun show(@PathVariable("user_id") userId: Long):
              ResponseEntity<UserVO> {
      val user = userService.retrieveUserData(userId)
      return ResponseEntity.ok(userAssembler.toUserVO(user))
    }

    @GetMapping
    @RequestMapping("/details")
    fun echoDetails(request: HttpServletRequest): ResponseEntity<UserVO>{
      val user = userRepository.findByUsername
                  (request.userPrincipal.name) as User
      return ResponseEntity.ok(userAssembler.toUserVO(user))
    }

    @GetMapping
    fun index(request: HttpServletRequest): ResponseEntity<UserListVO> {
      val user = userRepository.findByUsername
                  (request.userPrincipal.name) as User
      val users = userService.listUsers(user)

      return ResponseEntity.ok(userAssembler.toUserListVO(users))
    }

    @PutMapping
    fun update(@RequestBody updateDetails: User,
        request: HttpServletRequest): ResponseEntity<UserVO> {
      val currentUser = userRepository.findByUsername
                          (request.userPrincipal.name)
      userService.updateUserStatus(currentUser as User, updateDetails)
      return ResponseEntity.ok(userAssembler.toUserVO(currentUser))
    }
}
```

Now we are going to create controllers to handle message resources and conversation resources. These will be `MessageController` and `ConversationController`, respectively. Before creating the controllers, let's assemblers that will be used to assemble value objects from JPA entities. The following is the `MessageAssembler`:

```
package com.example.messenger.api.components

import com.example.messenger.api.helpers.objects.MessageVO
import com.example.messenger.api.models.Message
import org.springframework.stereotype.Component

@Component
class MessageAssembler {
  fun toMessageVO(message: Message): MessageVO {
    return MessageVO(message.id, message.sender?.id,
                     message.recipient?.id, message.conversation?.id,
                     message.body, message.createdAt.toString())
  }
}
```

And now, let's create the `ConversationAssembler`, as follows:

```
package com.example.messenger.api.components

import com.example.messenger.api.helpers.objects.ConversationListVO
import com.example.messenger.api.helpers.objects.ConversationVO
import com.example.messenger.api.helpers.objects.MessageVO
import com.example.messenger.api.models.Conversation
import com.example.messenger.api.services.ConversationServiceImpl
import org.springframework.stereotype.Component


@Component
class ConversationAssembler(val conversationService:
                            ConversationServiceImpl,
                            val messageAssembler: MessageAssembler) {

  fun toConversationVO(conversation: Conversation, userId: Long):
ConversationVO {
    val conversationMessages: ArrayList<MessageVO> = ArrayList()
    conversation.messages.mapTo(conversationMessages) {
      messageAssembler.toMessageVO(it)
    }
    return ConversationVO(conversation.id, conversationService
                          .nameSecondParty(conversation, userId),
                          conversationMessages)
  }
```

```
    fun toConversationListVO(conversations: ArrayList<Conversation>,
                             userId: Long): ConversationListVO {
      val conversationVOList = conversations.map { toConversationVO(it,
                                                   userId) }
      return  ConversationListVO(conversationVOList)
    }
  }
```

All is in place for `MessageController` and `ConversationController`. For our simple messenger app, we only need to have a message creation action for `MessageController`. The following is `MessageController` with the message creation action, `create`:

```
package com.example.messenger.api.controllers

import com.example.messenger.api.components.MessageAssembler
import com.example.messenger.api.helpers.objects.MessageVO
import com.example.messenger.api.models.User
import com.example.messenger.api.repositories.UserRepository
import com.example.messenger.api.services.MessageServiceImpl
import org.springframework.http.ResponseEntity
import org.springframework.web.bind.annotation.*
import javax.servlet.http.HttpServletRequest

@RestController
@RequestMapping("/messages")
class MessageController(val messageService: MessageServiceImpl,
                       val userRepository: UserRepository,
                       val messageAssembler: MessageAssembler) {

  @PostMapping
  fun create(@RequestBody messageDetails: MessageRequest,
             request: HttpServletRequest): ResponseEntity<MessageVO> {
    val principal = request.userPrincipal
    val sender = userRepository.findByUsername(principal.name) as User
    val message = messageService.sendMessage(sender,
                  messageDetails.recipientId, messageDetails.message)
    return ResponseEntity.ok(messageAssembler.toMessageVO(message))
  }

  data class MessageRequest(val recipientId: Long, val message: String)
}
```

Lastly, we must create `ConversationController`. We need only two endpoints: one to list all the active conversations of a user and the other to get the messages existing in a conversation thread. These endpoints will be catered to by the `list()` and `show()` actions, respectively. The following is the `ConversationController` class:

```
package com.example.messenger.api.controllers

import com.example.messenger.api.components.ConversationAssembler
import com.example.messenger.api.helpers.objects.ConversationListVO
import com.example.messenger.api.helpers.objects.ConversationVO
import com.example.messenger.api.models.User
import com.example.messenger.api.repositories.UserRepository
import com.example.messenger.api.services.ConversationServiceImpl
import org.springframework.http.ResponseEntity
import org.springframework.web.bind.annotation.*
import javax.servlet.http.HttpServletRequest

@RestController
@RequestMapping("/conversations")
class ConversationController(
  val conversationService: ConversationServiceImpl,
  val conversationAssembler: ConversationAssembler,
  val userRepository: UserRepository
) {

  @GetMapping
  fun list(request: HttpServletRequest): ResponseEntity<ConversationListVO>
{
    val user = userRepository.findByUsername(request
                  .userPrincipal.name) as User
    val conversations = conversationService.listUserConversations
                  (user.id)
    return ResponseEntity.ok(conversationAssembler
                  .toConversationListVO(conversations, user.id))
  }


  @GetMapping
  @RequestMapping("/{conversation_id}")
  fun show(@PathVariable(name = "conversation_id") conversationId: Long,
        request: HttpServletRequest): ResponseEntity<ConversationVO> {
    val user = userRepository.findByUsername(request
                       .userPrincipal.name) as User
    val conversationThread = conversationService.retrieveThread
                       (conversationId)
    return ResponseEntity.ok(conversationAssembler
                  .toConversationVO(conversationThread, user.id))
```

```
            }
        }
```

All is looking great! There's only one tiny problem. Remember, a user has an account status and it is possible for the account to be deactivated, right? In such a scenario, we, as API creators, will not want a deactivated user to be able to use our platform. As such, we have to come up with a way to prevent such a user from interacting with our API. There are a number of ways this can be done, but for this example we are going to use an interceptor. An interceptor intercepts an HTTP request and performs one or more operations on it before it continues down the request chain. Similar to assemblers, an interceptor is a component. We will call our interceptor that checks the validity of an account `AccountValidityInterceptor`. The following is the interceptor class (remember, it belongs in the `components` package):

```
package com.example.messenger.api.components

import com.example.messenger.api.exceptions.UserDeactivatedException
import com.example.messenger.api.models.User
import com.example.messenger.api.repositories.UserRepository
import org.springframework.stereotype.Component
import org.springframework.web.servlet.handler.HandlerInterceptorAdapter
import java.security.Principal
import javax.servlet.http.HttpServletRequest
import javax.servlet.http.HttpServletResponse

@Component
class AccountValidityInterceptor(val userRepository: UserRepository) :
        HandlerInterceptorAdapter() {

  @Throws(UserDeactivatedException::class)
  override fun preHandle(request: HttpServletRequest,
          response: HttpServletResponse, handler: Any?): Boolean {
    val principal: Principal? = request.userPrincipal

    if (principal != null) {
      val user = userRepository.findByUsername(principal.name)
                as User

      if (user.accountStatus == "deactivated") {
        throw UserDeactivatedException("The account of this user has
                                      been deactivated.")
      }
    }
    return super.preHandle(request, response, handler)
  }
}
```

The `AccountValidityInterceptor` class overrides the `preHandle()` function of its super class. This function will be called to carry out some operations prior to the routing of the request to its necessary controller action. After the creation of an interceptor, the interceptor must be registered with the Spring application. This configuration can be done using a `WebMvcConfigurer`. Add an `AppConfig` file to the `config` package in the project. Input the following code within the file:

```
package com.example.messenger.api.config

import com.example.messenger.api.components.AccountValidityInterceptor
import org.springframework.beans.factory.annotation.Autowired
import org.springframework.context.annotation.Configuration
import org.springframework.web.servlet.config.annotation.InterceptorRegistry
import org.springframework.web.servlet.config.annotation.WebMvcConfigurer

@Configuration
class AppConfig : WebMvcConfigurer {

  @Autowired
  lateinit var accountValidityInterceptor: AccountValidityInterceptor

  override fun addInterceptors(registry: InterceptorRegistry) {
    registry.addInterceptor(accountValidityInterceptor)
    super.addInterceptors(registry)
  }
}
```

`AppConfig` is a subclass of `WebMvcConfigurer` and overrides the `addInterceptor(InterceptorRegistry)` function in its superclass. `accountValidityInterceptor` is added to the interceptor registry with `registry.addInterceptor()`.

We are now done with all the code required to provide web resources to the messenger Android application. We must now deploy this code to a remote server.

Deploying the messenger API to AWS

Deploying a Spring Boot app to **Amazon Web Services** (**AWS**) is a straightforward and enjoyable process. The deployment procedure can be done well within 10 minutes. In this section, you will learn how to deploy applications based on Spring to AWS. Before application deployment, we must set up a PostgreSQL database on AWS that the application will connect to.

Setting up PostgreSQL on AWS

The first thing you must do is create an AWS account. Go ahead and create one now by following this link: `https://portal.aws.amazon.com/billing/signup#/start` . Once you have signed up, login into the AWS console and head over to the Amazon **Relational Database Service** (**RDS**) (from the navigation bar, click **Services** | **Database** | **RDS**). Once taken to the RDS dashboard, click on **Get Started Now**:

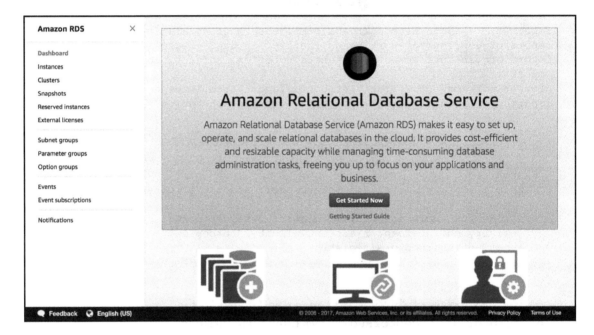

You will be navigated to the **Launch DB instance** web page. Here you will need to make some selections pertaining to the DB setup. Select **PostgreSQL** as the DB engine to be used:

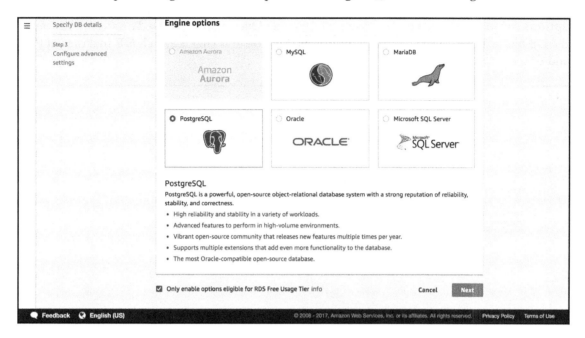

Ensure to check the **Only enable options eligible for RDS Free Usage Tier** checkbox. Navigate to the next set of setup procedures by clicking **Next**. On the next page, leave the instance specifications as is and input the necessary DB settings. Input a DB instance name, master username, and master password. We used `messenger-api` as our DB instance name, you may choose to use another name of your choice. Regardless your choice, ensure that you take note of all inputs you make. Once you are done entering the necessary, continue to the next screen.

You will be navigated to the **Configure advanced settings** screen. In the **Network & Security** section, ensure to enable public accessibility and select the **Create new VPC security group** option under **VPC Security Groups**. Scroll down to the **Database options** section of the screen and enter a DB name. Once again, we used `MessengerDB` as the DB name. Leave the remaining options as they are and click **Launch DB instance** at the end of the web page:

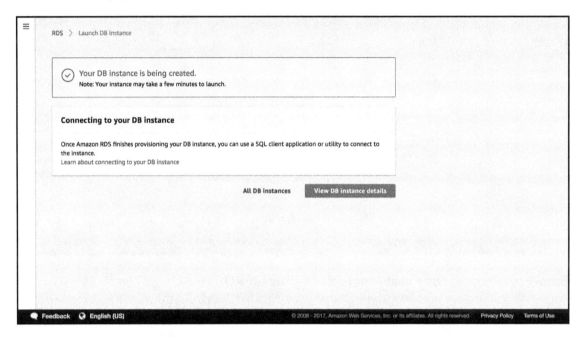

Your DB instance will be created by AWS. The creation process may take up to 10 minutes so it may be a good idea to take a coffee break at this point.

After waiting a little while, click **View DB instance details**. This will take you to a page where you can see detailed information on your just deployed DB instance. Scroll to the **Connect** section of the page to view the connection details of the DB instance:

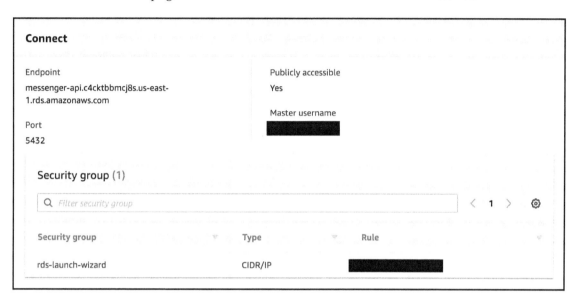

We need these details to successfully connect to the `MessengerDB` on this PostgreSQL DB instance. To enable `messenger-api` to connect to `MessengerDB`, you must edit the `spring.datasource.url`, `spring.datasource.username`, and `spring.datasource.password` properties in the `application.properties` file. After doing this, `application.properties` should look similar to the following:

```
spring.jpa.generate-ddl=true
spring.jpa.hibernate.ddl-auto=create-drop
spring.datasource.url=jdbc:postgresql://<endpoint>/MessengerDB
spring.datasource.username=<master_username>
spring.datasource.password=<password>
```

The final thing we will do is deploy the messenger API to an Amazon EC2 instance.

Deploying the messenger API to Amazon Elastic Beanstalk

Deploying an application to AWS is simple as well. Navigate to the AWS console and select **Services | Compute | Elastic Beanstalk.** Once in the Elastic Beanstalk Management Console, click on **Create New Application**. Once taken to the **Create Application** page, you will be asked to provide a name and description for the new application to be created. Name the app `messenger-api` and proceed to the next screen. You will be taken to a screen in which you are prompted to create a new environment:

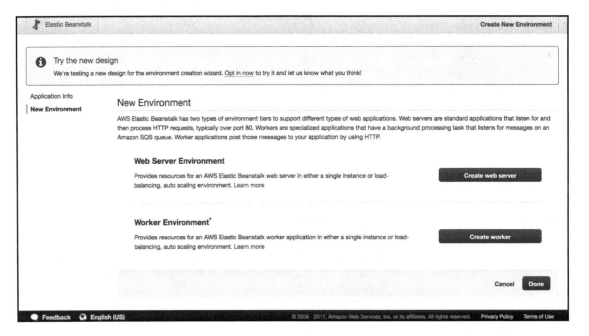

Create a web server environment on this screen. Next up, you will need to configure the environment type. Select the **Tomcat** predefined configuration and change the environment type to **Single instance**:

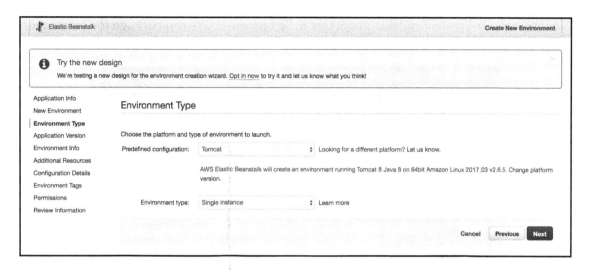

Continue to the next screen when you are ready. On the next screen, you will need to select a source for your application. Choose **Upload your own**:

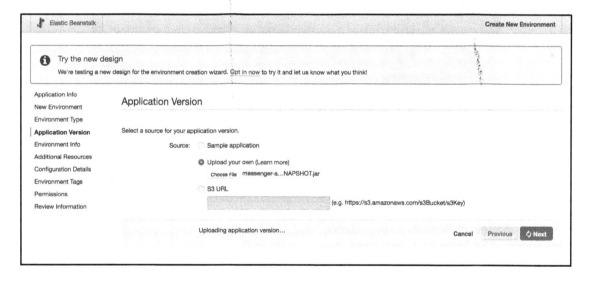

Now we need to create a suitable project jar to be uploaded. We can package the messenger API into a jar with the help of Maven:

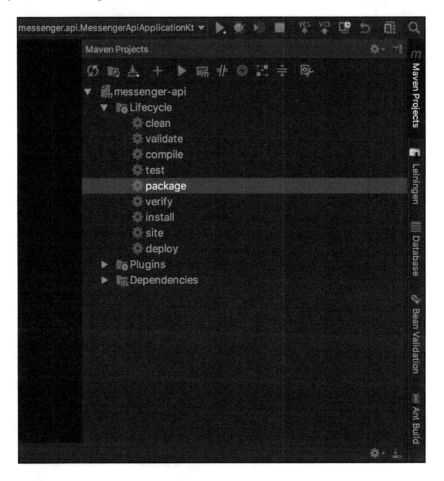

Click the **Maven Projects** button to the right of your project IDE screen and select **messenger-api** | **Lifecycle** | **package**. A project jar will be packaged and stored in the target directory of the project.

Now go back to AWS and choose this jar file as the source file to be uploaded. Leave other properties as they are and click **Next**. You may have to wait for a few minutes while the packaged jar is uploaded. Once the upload is done, you will be taken to a new screen where your environment information is presented to you. Proceed through the next few screens by clicking **Next** until you are presented with a **Configuration Details** screen. Change the instance type to t2.micro.

Proceed through the next screens until you reach the **Review Information** section. Scroll through this page until you reach the **Environment Info** section:

Environment Info	
Environment name	messengerApi-env
Environment URL	http://messengerapi-env.us-east-1.elasticbeanstalk.com
Description	REST API for the Messenger app

Your environment URL will differ. Take note of this information as you will need it later on. Scroll to the bottom of the web page and click **Launch**. Elastic Beanstalk will launch your new environment.

Once the launch finishes, you are good to go. You have successfully deployed the `messenger-api` to AWS.

Summary

In this chapter, we explored how to utilize Kotlin to build a Spring Boot REST application programming interface. In the process of doing so, we learned the basics of designing systems. We expressed the behavior of the messenger API system with a state diagram, and learned how to properly interpret the information represented in a state diagram. We went one step further by creating an E-R diagram to give a detailed diagrammatic representation of system entities and their relationships.

In addition, we learned how to set up PostgreSQL on a local machine and create a new PostgreSQL database. We also explored how to build a microservice with Spring Boot 2.0, connect the microservice to a database, and interact with data existing in a database with Spring Data.

Besides all this, we learned how to properly secure a RESTful Spring Boot web application with Spring Security and **JSON Web Tokens** (**JWTs**). We created custom Spring Security configurations to facilitate user authentication with the help of JWTs and created custom filters for the authentication process. Lastly, we learned how to deploy a Spring Boot application to AWS.

In the next chapter, you will explore Kotlin further in the Android domain by building the Android messenger application.

5
Building the Messenger Android App – Part I

In the previous chapter, we began building the messenger application by designing and implementing a REST application programming interface that the client messenger application will communicate with. Over the course of implementing the backend API, we covered many things, such as working with Spring Boot, RESTful application programming interfaces and how they work, creating databases with PostgreSQL, and deploying Spring Boot web applications to AWS, to name a few.

In this chapter, we will go one step further in our application development journey by implementing the Android Messenger application and integrating it with the RESTful API we created in Chapter 4, *Designing and Implementing the Messenger Backend with Spring Boot 2.0*. In the process of developing the Messenger Android app, we will learn a vast array of new topics, such as:

- Building MVP Android applications
- Server communication via HTTP
- Working with Retrofit
- Reactive programming
- Using token-based authentication in an Android app

Over the course of this chapter, you will learn firsthand how powerful Kotlin is in the Android application development domain. Let's dive into the development of the Messenger app.

Developing the Messenger app

First, we need to create a new Android Studio project for the application. Create a new Android Studio project with the name `Messenger` and the package name `com.example.messenger`. Feel free to take a look at `Chapter 1`, *The Fundamentals*, to refresh your memory on Android project creation. In the process of project setup, when asked to create a new launcher activity, name the activity `LoginActivity` and make it an empty activity.

Including project dependencies

Over the course of this chapter, we will make use of a number of external application dependencies. As such, it is important we include them in the project now. Open your module-level `build.gradle` file and add the following dependencies to it:

```
dependencies {
    implementation fileTree(dir: 'libs', include: ['*.jar'])
    implementation "org.jetbrains.kotlin:kotlin-stdlib-jre7
                    :$kotlin_version"
    implementation 'com.android.support:appcompat-v7:26.1.0'
    implementation 'com.android.support.constraint:constraint-layout:1.0.2'
    implementation 'com.android.support:recyclerview-v7:26.1.0'
    implementation 'com.android.support:design:26.1.0'

    implementation "android.arch.persistence.room:runtime:1.0.0-alpha9-1"
    implementation "android.arch.persistence.room:rxjava2:1.0.0-alpha9-1"
    implementation 'com.android.support:support-v4:26.1.0'
    implementation 'com.android.support:support-vector-drawable:26.1.0'
    annotationProcessor "android.arch.persistence.room:compiler
                    :1.0.0-alpha9-1"

    implementation "com.squareup.retrofit2:retrofit:2.3.0"
    implementation "com.squareup.retrofit2:adapter-rxjava2:2.3.0"
    implementation "com.squareup.retrofit2:converter-gson:2.3.0"
    implementation "io.reactivex.rxjava2:rxandroid:2.0.1"

    implementation 'com.github.stfalcon:chatkit:0.2.2'

    testImplementation 'junit:junit:4.12'
    androidTestImplementation 'com.android.support.test:runner:1.0.1'
    androidTestImplementation 'com.android.support.test.espresso
                    :espresso-core:3.0.1'
}
```

Ensure that no conflicting Android support library versions exist in the `build.gradle` file. Now modify the `build.gradle` project file to include the `jcenter` and Google repositories as well as the Android build tools dependency:

Top-level build file where you can add configuration options common to all sub-projects/modules:

```
buildscript {
    ext.kotlin_version = '1.1.4-3'
    repositories {
        google()
        jcenter()
    }
    dependencies {
        classpath 'com.android.tools.build:gradle:3.0.0-alpha9'
        classpath "org.jetbrains.kotlin:kotlin-gradle-plugin:$kotlin_version"
    }
}

allprojects {
    repositories {
        google()
        jcenter()
    }
}

task clean(type: Delete) {
    delete rootProject.buildDir
}
```

Don't worry, about what the dependencies added for now, all will be revealed over the course of this chapter.

Developing the Login UI

Once the project is created, create a new package named `ui` in the `com.example.messenger` application source package. This package will hold all the user-interface-related classes and logic of the Android application. Create a `login` package within `ui`. As you may have guessed, this package will hold classes and logic pertaining to the user-login process. Go ahead and move `LoginActivity` to the `login` package. Having moved `LoginActivity`, our first order of business is to create a suitable layout for the login activity.

Locate the `activity_login.xml` layout resource file and change the following content:

```xml
<?xml version="1.0" encoding="utf-8"?>
<LinearLayout xmlns:android="http://schemas.android.com/apk/res/android"
    xmlns:tools="http://schemas.android.com/tools"
    android:layout_width="match_parent"
    android:layout_height="match_parent"
    tools:context=".ui.login.LoginActivity"
    android:orientation="vertical"
    android:paddingTop="32dp"
    android:paddingBottom="@dimen/default_margin"
    android:paddingStart="@dimen/default_padding"
    android:paddingEnd="@dimen/default_padding"
    android:gravity="center_horizontal">
    <EditText
        android:id="@+id/et_username"
        android:layout_width="match_parent"
        android:layout_height="wrap_content"
        android:inputType="text"
        android:hint="@string/username"/>
    <EditText
        android:id="@+id/et_password"
        android:layout_width="match_parent"
        android:layout_height="wrap_content"
        android:layout_marginTop="@dimen/default_margin"
        android:inputType="textPassword"
        android:hint="@string/password"/>
    <Button
        android:id="@+id/btn_login"
        android:layout_width="wrap_content"
        android:layout_height="wrap_content"
        android:layout_marginTop="@dimen/default_margin"
        android:text="@string/login"/>
    <Button
        android:id="@+id/btn_sign_up"
        android:layout_width="wrap_content"
        android:layout_height="wrap_content"
        android:layout_marginTop="@dimen/default_margin"
        android:background="@android:color/transparent"
        android:text="@string/sign_up_solicitation"/>
    <ProgressBar
        android:id="@+id/progress_bar"
        android:layout_width="wrap_content"
        android:layout_height="wrap_content"
        android:visibility="gone"/>
</LinearLayout>
```

There are string and dimension resources that we have made use of here that we have not yet created in the appropriate .xml resource files. We must add these resources at this juncture. While we are at it, we will also include resources that we will require later on in the application development phase so as to eliminate the need to jump back and forth between program files and resource files. Open the project's string resource file (strings.xml) and ensure the following resources are added to it:

```xml
<resources>
  <string name="app_name">Messenger</string>
  <string name="username">Username</string>
  <string name="password">Password</string>
  <string name="login">Login</string>
  <string name="sign_up_solicitation">
    Don\'t have an account? Sign up!
  </string>
  <string name="sign_up">Sign up</string>
  <string name="phone_number">Phone number</string>
  <string name="action_settings">settings</string>
  <string name="hint_enter_a_message">Type a message...</string>

  <!-- Account settings -->
  <string name="title_activity_settings">Settings</string>
  <string name="pref_header_account">Account</string>
  <string name="action_logout">logout</string>
</resources>
```

Now create a dimensions resource file (dimens.xml) and add the following dimension resources to it:

```xml
<?xml version="1.0" encoding="utf-8"?>
<resources>
  <dimen name="default_margin">16dp</dimen>
  <dimen name="default_padding">16dp</dimen>
</resources>
```

Now that the necessary project resources have been added, navigate back to
`activity_login.xml` and toggle the design preview screen to view the layout that has
been created:

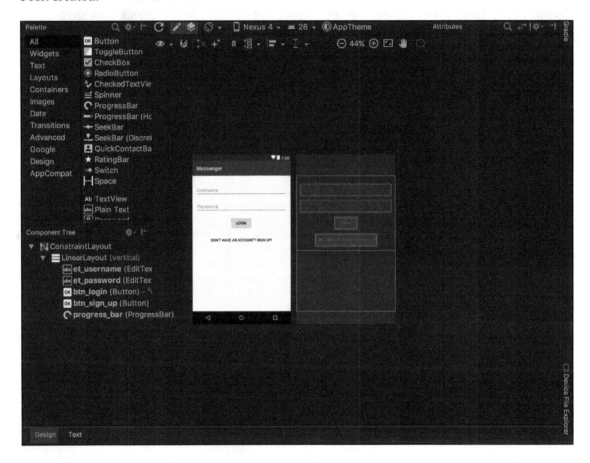

The layout is simple but functional, which is perfect for this simple messenger application
we are building.

Creating the login view

Now we have to work on `LoginActivity`. As we are building this application utilizing the MVP pattern, `LoginActivity` is effectively a view. Obviously, `LoginActivity` is quite different from any generic view. It is a view that is concerned with login procedures. We can identify a set of necessary behaviors that a view presenting a login interface to a user must possess. Such behaviors are:

- It must show a progress bar to a user when the login is in progress
- It must be capable of hiding the progress bar when the need arises
- It must be able to show appropriate field errors to users when encountered
- It must be able to navigate the user to his home screen
- It must be able to navigate an unregistered user to a signup screen

Having identified the preceding behaviors, we must ensure that `LoginActivity`—as a login view—exhibits such behaviors. A perfect way to do this is to utilize an interface. Create a `LoginView` interface in the `login` package containing the following content:

```
package com.example.messenger.ui.login

interface LoginView {
    fun showProgress()
    fun hideProgress()
    fun setUsernameError()
    fun setPasswordError()
    fun navigateToSignUp()
    fun navigateToHome()
}
```

So far, so good with `LoginView`. There are a few issues with our interface though. A `LoginView` must have the ability to bind its layout views to appropriate object representations. In addition, a `LoginView` must be able to provide feedback to the user if an authentication error occurs. You may be thinking that neither of these two behaviors should be distinct to a `LoginView`. You are right. All views should have the ability to bind layout elements to programmatic objects. In addition, a signup view should also be able to provide the user with some kind of feedback if there is a problem during authentication.

We will create two distinct interfaces to enforce these behaviors. We will name the first interface `BaseView`. **Create a** `base` **package in** `com.example.messenger.ui` **and add an** interface named `BaseView` to the package with the following content:

```
package com.example.messenger.ui.base

import android.content.Context

interface BaseView {
  fun bindViews()
  fun getContext(): Context
}
```

The `BaseView` interface enforces that an implementing class declares `bindViews()` and `getContext()` functions for view bindings and context retrievals, respectively.

Now create an `auth` package in `com.example.messenger.ui` and add an interface named `AuthView` to the package with the following content:

```
package com.example.messenger.ui.auth

interface AuthView {
  fun showAuthError()
}
```

Fantastic job! Now go back to the `LoginView` interface and ensure that it extends `BaseView` and `AuthView`, as follows:

```
package com.example.messenger.ui.login

import com.example.messenger.ui.auth.AuthView
import com.example.messenger.ui.base.BaseView

interface LoginView : BaseView, AuthView {
  fun showProgress()
  fun hideProgress()
  fun setUsernameError()
  fun setPasswordError()
  fun navigateToSignUp()
  fun navigateToHome()
}
```

By declaring the `LoginView` interface as an extension of `BaseView` and `AuthView`, we ensure that every class that implements `LoginView` must declare the `bindViews()`, `getContext()`, and `showAuthError()` functions in addition to those declared in `LoginView`. It is important to note that any class that implements `LoginView` is effectively of the `LoginView`, `BaseView`, and `AuthView` type. The characteristic of a class possessing many types is known as polymorphism.

Having set up the `LoginView`, we can go ahead and work on `LoginActivity`. Firstly we will create `LoginActivity` to implement the methods declared in `BaseView` and `AuthView` after which we will add the methods specific to a `LoginView`. `LoginActivity` is shown in the following code:

```
package com.example.messenger.ui.login

import android.content.Context
import android.content.Intent
import android.support.v7.app.AppCompatActivity
import android.os.Bundle
import android.view.View
import android.widget.Button
import android.widget.EditText
import android.widget.ProgressBar
import android.widget.Toast
import com.example.messenger.R

class LoginActivity : AppCompatActivity(), LoginView, View.OnClickListener
{

  private lateinit var etUsername: EditText
  private lateinit var etPassword: EditText
  private lateinit var btnLogin: Button
  private lateinit var btnSignUp: Button
  private lateinit var progressBar: ProgressBar

  override fun onCreate(savedInstanceState: Bundle?) {
    super.onCreate(savedInstanceState)
    setContentView(R.layout.activity_login)

    bindViews()
  }
```

Binds layout view object references to view elements when invoked:

```
override fun bindViews() {
  etUsername = findViewById(R.id.et_username)
  etPassword = findViewById(R.id.et_password)
  btnLogin = findViewById(R.id.btn_login)
  btnSignUp = findViewById(R.id.btn_sign_up)
  progressBar = findViewById(R.id.progress_bar)
  btnLogin.setOnClickListener(this)
  btnSignUp.setOnClickListener(this)
}

/**
  * Shows an appropriate Authentication error message when invoked.
*/
override fun showAuthError() {
  Toast.makeText(this, "Invalid username and password combination.",
Toast.LENGTH_LONG).show()
}

override fun onClick(view: View) {
}

override fun getContext(): Context {
  return this
}
}
```

So far so good. We have successfully implemented `BaseView` and `AuthView` methods in `LoginActivity`. We still must work on the `LoginView` specific methods `showProgress()`, `hideProgress()`, `setUsernameError()`, `setPasswordError()`, `navigateToSignUp()` and `navigateToHome()`. The required implementation of these methods is given below. Go ahead and add them to `LoginActivity`.

```
override fun hideProgress() {
  progressBar.visibility = View.GONE
}

override fun showProgress() {
  progressBar.visibility = View.VISIBLE
}

override fun setUsernameError() {
  etUsername.error = "Username field cannot be empty"
}

override fun setPasswordError() {
```

```
    etPassword.error = "Password field cannot be empty"
  }

  override fun navigateToSignUp() {
  }

  override fun navigateToHome() {
  }
```

Adding all the defined methods earlier, you have made the `LoginActivity` class to implement `LoginView` as well as the `View.OnClickListener` interface. As such, `LoginActivity` provides implementations for functions declared within these interfaces. Notice how the current instance of the `LoginActivity` is passed as an argument to `btnLogin.setOnClickListener()` via `this`. We can do this because we have declared `LoginActivity` to implement the `View.OnClickListener` interface. As such, `LoginActivity` is a valid `View.OnClickListener` instance (this is a perfect example of polymorphism at work).

Now that we have done some reasonable work on the login view, we must create an appropriate model to handle login logic. We must also create the necessary services and data repositories that this model will communicate with. We will first work on the required services and then develop the necessary data repositories before we build the interactor.

Creating the Messenger API service and data repositories

A critical thing that we must consider before getting too deep in our application development process is data storage. We must ask ourselves two very important questions: Where will data be stored, and how will the stored data be accessed?

Pertaining to the location of data storage, data will be stored both locally (on the Android device) and remotely (on the Messenger API). The answer to the second question is similarly straightforward. To access stored data, we need to create suitable models, services, and repositories to facilitate data retrieval.

Storing data locally with SharedPreferences

First and foremost, we need to take care of local storage. As this is a simple application, we do not need to store much data locally. The only data we need to store on the device are access tokens and user details. We will be using `SharedPreferences` to do this.

First things first, create a `data` package within the application's `source` package. We previously identified that we are going to be working with data stored locally and remotely. Hence, create two additional packages within `data`. Name the first `local` and the second `remote`. Similar to the approach that we used for the *Tetris* application that we created in Chapter 2, *Building an Android Application – Tetris* and Chapter 3, *Implementing Tetris Logic and Functionality*, we will be using an `AppPreferences` class to persist data locally. Create an `AppPreferences` class within `local` and populate it with the following content:

```
package com.example.messenger.data.local

import android.content.Context
import android.content.SharedPreferences
import com.example.messenger.data.vo.UserVO

class AppPreferences private constructor() {

  private lateinit var preferences: SharedPreferences

  companion object {
    private val PREFERENCE_FILE_NAME = "APP_PREFERENCES"

    fun create(context: Context): AppPreferences {
      val appPreferences = AppPreferences()
      appPreferences.preferences = context
      .getSharedPreferences(PREFERENCE_FILE_NAME, 0)
        return appPreferences
      }
    }

  val accessToken: String?
    get() = preferences.getString("ACCESS_TOKEN", null)

  fun storeAccessToken(accessToken: String) {
    preferences.edit().putString("ACCESS_TOKEN", accessToken).apply()
  }

  val userDetails: UserVO
  get(): UserVO {
```

Returns an instance of `UserVO` containing appropriate user details:

```
    return UserVO(
      preferences.getLong("ID", 0),
      preferences.getString("USERNAME", null),
      preferences.getString("PHONE_NUMBER", null),
      preferences.getString("STATUS", null),
```

```
            preferences.getString("CREATED_AT", null)
        )
    }
```

Stores user details passed in `UserVO` to the application's `SharedPreferences` file:

```
    fun storeUserDetails(user: UserVO) {
        val editor: SharedPreferences.Editor = preferences.edit()

        editor.putLong("ID", user.id).apply()
        editor.putString("USERNAME", user.username).apply()
        editor.putString("PHONE_NUMBER", user.phoneNumber).apply()
        editor.putString("STATUS", user.status).apply()
        editor.putString("CREATED_AT", user.createdAt).apply()
    }

    fun clear() {
        val editor: SharedPreferences.Editor = preferences.edit()
        editor.clear()
        editor.apply()
    }
}
```

In `AppPreferences`, we defined the `storeAccessToken(String)`, `storeUserDetails(UserVO)`, and `clear()` functions. `storeAccessToken(String)` will be used to store an access token retrieved from the remote server to the local preferences file. `storeUserDetails(UserVO)` takes a user value object (a data object that contains user information) as its only argument and stores the information contained in the value object to the preferences file. The `clear()` method, as the name implies, clears all values that have been stored in the preferences file. The `AppPreferences` instance also has `accessToken` and `userDetails` properties, each having specialized getter functions to retrieve their appropriate values. In addition to this the functions and properties defined in `AppPreferences`, we also created a companion object possessing a single `create(Context)` function. The `create()` method, as the name implies, creates and returns a new `AppPreferences` instance for use. We made the primary constructor of `AppPreferences` private as we require that any class utilizing `AppPreferences` make use of `create()` for the instantiation of `AppPreferences`.

Creating value objects

Similar to what we did when creating the messenger backend, we need to create value objects to model common types of data we will be handling across the application. Create a vo package in the data package. The value objects we are creating are already familiar to you. In fact, they are exactly the same as those we created during the development of the API. We are going to create ConversationListVO, ConversationVO, UserListVO, UserVO, and MessageVO. Create Kotlin files to hold each of these value objects in the vo package. Before creating any list value object data models, we have to create the basic models. These models are UserVO, MessageVO, and ConversationVO.

Create a UserVO data class, as follows:

```
package com.example.messenger.data.vo

data class UserVO(
  val id: Long,
  val username: String,
  val phoneNumber: String,
  val status: String,
  val createdAt: String
)
```

As we have created value objects in the past, the previous code doesn't need much explaining. Add MessageVO to your MessageVO.kt file, as follows:

```
package com.example.messenger.data.vo

data class MessageVO(
  val id: Long,
  val senderId: Long,
  val recipientId: Long,
  val conversationId: Long,
  val body: String,
  val createdAt: String
)
```

Now create a ConversationVo data class in ConversationVO.kt, as follows:

```
package com.example.messenger.data.vo

data class ConversationVO(
  val conversationId: Long,
  val secondPartyUsername: String,
  val messages: ArrayList<MessageVO>
)
```

Having created the basic value objects, let's create `ConversationListVO` and `UserListVO`, shall we? `ConversationListVO` is as follows:

```
package com.example.messenger.data.vo

data class ConversationListVO(
  val conversations: List<ConversationVO>
)
```

The `ConversationListVO` data class has a single `conversations` property of the `List` type that can only contain elements of the `ConversationVO` type. The `UserListVO` data class is similar to `ConversationListVO`, with the exception that it has a user's property, which can only contain elements of the `UserVO` type instead of a `conversations` property. The following is the `UserListVO` data class:

```
package com.example.messenger.data.vo

data class UserListVO(
  val users: List<UserVO>
)
```

Retrieving remote data

We have already established that important data necessary for the functioning the messenger Android App will be stored remotely on the messenger backend. It is imperative that we have an efficient means by which our Android application can access the data held by the backend. To do this, the Messenger application needs to be able to communicate with the API via HTTP.

Communicating with a remote server

There are a number of ways that communication with a remote server can be achieved in Android. Common networking libraries used across the Android community are **Retrofit**, **OkHttp**, and **Volley**. Each of these libraries has its advantages and drawbacks. We will be making use of Retrofit in this project, but for the sake of knowledge, we will take a look at how to communicate with a remote server using OkHttp.

Communicating with servers using OkHttp

OkHttp is an efficient and easy-to-use HTTP client. It supports both synchronous and asynchronous network calls. Using OkHttp on Android is easy. Simply add its dependency to a project's module-level `build.gradle` file:

```
implementation 'com.squareup.okhttp3:okhttp:3.9.0'
```

Sending requests to a server with OkHttp

As previously stated, OkHttp's APIs were built with ease of use in mind. As a consequence, sending requests via OkHttp is quick and hassle-free. The following is a `post(String, String)` method that takes a URL and JSON request body as its arguments and sends a POST request to the specified URL with the JSON body:

```kotlin
fun post(url: String, json: String): String {
    val mediaType: MediaType = MediaType.parse("application/json;
                                                charset=utf-8")
    val client:OkHttpClient = OkHttpClient()
    val body: RequestBody = RequestBody.create(mediaType, json)

    val request: Request = Request.Builder()
                                .url(url)
                                .post(body)
                                .build()

    val response: Response = client.newCall(request).execute()
    return response.body().string()
}
```

Using the preceding function is straightforward. Invoke it with appropriate values, as you would any other function:

```kotlin
val fullName: String = "John Wayne"
val response = post("http://example.com", "{ \"full_name\": $fullName")

println(response)
```

Easy, right? Glad you agree. Communicating with a remote server is fun with OkHttp but using Retrofit to do this is even more fun. We are almost ready to work with Retrofit, but, before we explore Retrofit, it's a good idea to properly model the data we will be sending in our HTTP requests.

Modeling request data

We will make use of data classes to model the HTTP request data we wish to send to our API. Go ahead and create a `request` package in the `remote` package. There are four obvious requests that contain data payloads, which we will be sending to the API. These are login requests, message requests, status update requests, and requests containing user data. These four requests will be modeled by `LoginRequestObject`, `MessageRequestObject`, `StatusUpdateRequestObject`, and `UserRequest` objects, respectively.

The following code snippet shows the `LoginRequestObject` data class. Go ahead and add it to the `request` package and do the same for other request objects that follow:

```
package com.example.messenger.data.remote.request

data class LoginRequestObject(
  val username: String,
  val password: String
)
```

The `LoginRequestObject` data class possesses the `username` and `password` properties because these are the credentials that need to be supplied to the login endpoint of the API. The `MessageRequestObject` data class is as follows:

```
package com.example.messenger.data.remote.request

data class MessageRequestObject(val recipientId: Long, val message: String)
```

`MessageRequestObject` possesses two properties as well. These are `recipientId`—the ID of a user receiving a message—and `message` —the body of the message being sent:

```
package com.example.messenger.data.remote.request

data class StatusUpdateRequestObject(val status: String)
```

The `StatusUpdateRequestObject` data class has a single `status` property. As the name implies, this is the status that a user wants to update their current status message to:

```
package com.example.messenger.data.remote.request

data class UserRequestObject(
  val username: String,
  val password: String,
  val phoneNumber: String = ""
)
```

`UserRequestObject` is similar to `LoginRequestObject` with the exception that it contains an additional `phoneNumber` property. This request object has varying use cases, such as to contain user signup data being sent to the API.

Having created the necessary request objects, we can go ahead and create the actual `MessengerApiService`.

Creating the Messenger API service

It's time for us to create a service that performs the all-important job of communicating with the Messenger API we created in `Chapter 4`, *Designing and Implementing the Messenger Backend with Spring Boot 2.0*. We will be making use of Retrofit and Retrofit's RxJava adapter to create this service. Retrofit is a type-safe HTTP client for Android and Java built by Square Inc., and RxJava is an open source implementation of ReactiveX written in and for Java.

We added Retrofit to our Android project at the beginning of this chapter with the following line:

```
implementation "com.squareup.retrofit2:retrofit:2.3.0"
```

We also added Retrofit's RxJava adapter dependency to our module-level `build.gradle` script, as follows:

```
implementation "com.squareup.retrofit2:adapter-rxjava2:2.3.0"
```

The first step in creating a service with Retrofit is to define an interface that describes your HTTP API. Create a `service` package within your application `source` package and add the `MessengerApiService` interface, as follows:

```
package com.example.messenger.service

import com.example.messenger.data.remote.request.LoginRequestObject
import com.example.messenger.data.remote.request.MessageRequestObject
import com.example.messenger.data.remote.request.StatusUpdateRequestObject
import com.example.messenger.data.remote.request.UserRequestObject
import com.example.messenger.data.vo.*
import io.reactivex.Observable
import okhttp3.ResponseBody
import retrofit2.Retrofit
import retrofit2.adapter.rxjava2.RxJava2CallAdapterFactory
import retrofit2.converter.gson.GsonConverterFactory
import retrofit2.http.*

interface MessengerApiService {
```

```kotlin
@POST("login")
@Headers("Content-Type: application/json")
fun login(@Body user: LoginRequestObject):
        Observable<retrofit2.Response<ResponseBody>>

@POST("users/registrations")
fun createUser(@Body user: UserRequestObject): Observable<UserVO>

@GET("users")
fun listUsers(@Header("Authorization") authorization: String):
            Observable<UserListVO>

@PUT("users")
fun updateUserStatus(
  @Body request: StatusUpdateRequestObject,
  @Header("Authorization") authorization: String): Observable<UserVO>

@GET("users/{userId}")
fun showUser(
  @Path("userId") userId: Long,
  @Header("Authorization") authorization: String): Observable<UserVO>

@GET("users/details")
fun echoDetails(@Header("Authorization") authorization: String):
Observable<UserVO>


@POST("messages")
fun createMessage(
  @Body messageRequestObject: MessageRequestObject,
  @Header("Authorization") authorization: String): Observable<MessageVO>

@GET("conversations")
fun listConversations(@Header("Authorization") authorization: String):
                      Observable<ConversationListVO>

@GET("conversations/{conversationId}")
fun showConversation(
  @Path("conversationId") conversationId: Long,
  @Header("Authorization") authorization:
String):Observable<ConversationVO>
}
```

As can be observed from the preceding code snippet, Retrofit relies on the use of annotations to properly describe HTTP requests to be sent. Take the following snippet, for example:

```
@POST("login")
@Headers("Content-Type: application/json")
fun login(@Body user: LoginRequestObject):
Observable<retrofit2.Response<ResponseBody>>
```

The `@POST` annotation tells Retrofit that this function describes an HTTP POST request that is mapped to the `/login` path. The `@Headers` annotation is used to specify the headers of the HTTP request. In the HTTP request described in the preceding code snippet, the `Content-Type` header has been set to `application/json`. Hence, the content being sent by this request is JSON.

The `@Body` annotation specifies that the `user` argument passed to `login()` contains the data of the JSON request body to be sent to the API. `user` is of the `LoginRequestObject` type (we previously created this request object). Lastly, the function is declared to return an `Observable` object containing a `retrofit2.Response` object.

Besides the `@POST`, `@Headers`, and `@Body` annotations, we made use of `@GET`, `@PUT`, `@Path`, and `@Header`. `@GET` and `@PUT` are used to specify `GET` and `PUT` requests, respectively. The `@Path` annotation is used to declare a value as a path argument of the HTTP request being sent. Take the following `showUser()` function, for example:

```
@GET("users/{userId}")
fun showUser(
  @Path("userId") userId: Long,
  @Header("Authorization") authorization: String): Observable<UserVO>
```

`showUser` is a function that describes a GET request with the `users/{userId}` path. `{userId}` is not actually part of the HTTP request path. Retrofit will replace `{userId}` with the value passed to the `userId` argument of `showUser()`. Notice how `userId` is annotated with `@Path("userId")`. This lets retrofit know that `userId` holds a value that should be placed where `{userId}` is located in the HTTP request URL path.

`@Header` is similar to `@Headers`, with the exception that it is used to specify a single header key-value pair in an HTTP request to be sent. Annotating authorization with `@Header("Authorization")` sets the `Authorization` header of the HTTP request sent to the value held within authorization.

Now that we have created an appropriate `MessengerApiService` interface to model the HTTP API that our application will communicate with, we need to be able to retrieve an instance of this service. We can easily do this by creating a `Factory` companion object that's in charge of the creation of `MessengerApiService` instances:

```
package com.example.messenger.service

import com.example.messenger.data.remote.request.LoginRequestObject
import com.example.messenger.data.remote.request.MessageRequestObject
import com.example.messenger.data.remote.request.StatusUpdateRequestObject
import com.example.messenger.data.remote.request.UserRequestObject
import com.example.messenger.data.vo.*
import io.reactivex.Observable
import okhttp3.ResponseBody
import retrofit2.Retrofit
import retrofit2.adapter.rxjava2.RxJava2CallAdapterFactory
import retrofit2.converter.gson.GsonConverterFactory
import retrofit2.http.*

interface MessengerApiService {
    ...
    ...
    companion object Factory {
        private var service: MessengerApiService? = null
```

It returns an instance of `MessengerApiService` when invoked. A new instance of `MessengerApiService` is created, if one has not been previously created

```
        fun getInstance(): MessengerApiService {
            if (service == null) {

                val retrofit = Retrofit.Builder()
                        .addCallAdapterFactory(RxJava2CallAdapterFactory.create())
                        .addConverterFactory(GsonConverterFactory.create())
                        .baseUrl("{AWS_URL}")
                        // replace AWS_URL with URL of AWS EC2
                        // instance deployed in the previous chapter
                        .build()

                service = retrofit.create(MessengerApiService::class.java)
            }

            return service as MessengerApiService
        }
    }
}
```

`Factory` possesses a single `getInstance()` function that builds and returns an instance of `MessengerApiService` when called. An instance of `Retrofit.Builder` is used to create the interface. We set the `CallAdapterFactory` in use to `RxJava2CallAdapterFactory` and we set the `ConverterFactory` in use to `GsonConverterFactory` (this handles JSON serialization and deserialization). Don't forget to replace `"{AWS_URL}"` with the URL of the Messenger API AWS EC2 instance deployed in `Chapter 4`, *Designing and Implementing the Messenger Backend with Spring Boot 2.0*.

After creating the `Retrofit.Builder()` instance successfully, we use it to create an instance of `MessengerApiService`, as follows:

```
service = retrofit.create(MessengerApiService::class.java)
```

Lastly, the service is returned for use by `getInstance()`. Regardless of the fact that we have created a suitable service to communicate with the Messenger API, it cannot be used to communicate with a network without specifying the necessary permissions in the `AndroidManifest`. Open the project's `AndroidManifest` and add the following two lines of code within the `<manifest></manifest>` tag:

```
<uses-permission android:name="android.permission.INTERNET" />
<uses-permission android:name="android.permission.ACCESS_NETWORK_STATE" />
```

Now that we have the messenger service ready to go, it is time that we create appropriate repositories to exploit this service.

Implementing data repositories

You are already familiar with repositories, so there is no need for an introduction to them. The repositories we are about to create are similar to those created for the Messenger API in `Chapter 4`, *Designing and Implementing the Messenger Backend with Spring Boot 2.0*. The only difference is that the data source for the repositories we are about to implement is a remote server, not a database residing on a host.

Create a `repository` package within the `remote` package. First and foremost, we are going to implement a user repository to retrieve data pertaining to application users. Add a `UserRepository` interface to the repository, as follows:

```
package com.example.messenger.data.remote.repository

import com.example.messenger.data.vo.UserListVO
import com.example.messenger.data.vo.UserVO
import io.reactivex.Observable
```

```
interface UserRepository {

  fun findById(id: Long): Observable<UserVO>
  fun all(): Observable<UserListVO>
  fun echoDetails(): Observable<UserVO>
}
```

As this is an interface, we need to create a class that implements the functions specified within UserRepository. We will name this class UserRepositoryImpl. Create a new UserRepositoryImpl within the repositories package, as follows:

```
package com.example.messenger.data.remote.repository

import android.content.Context
import com.example.messenger.service.MessengerApiService
import com.example.messenger.data.local.AppPreferences
import com.example.messenger.data.vo.UserListVO
import com.example.messenger.data.vo.UserVO
import io.reactivex.Observable

class UserRepositoryImpl(ctx: Context) : UserRepository {

  private val preferences: AppPreferences = AppPreferences.create(ctx)
  private val service: MessengerApiService =
MessengerApiService.getInstance()

  override fun findById(id: Long): Observable<UserVO> {
    return service.showUser(id, preferences.accessToken as String)
  }

  override fun all(): Observable<UserListVO> {
    return service.listUsers(preferences.accessToken as String)
  }

  override fun echoDetails(): Observable<UserVO> {
    return service.echoDetails(preferences.accessToken as String)
  }
}
```

The preceding UserRepositoryImpl class has two instance variables: preferences and service. The preferences variable is an instance of the AppPreferences class we created earlier and service is an instance of MessengerApiService retrieved by the getInstance() function defined in the Factory companion object in the MessengerApiService interface.

`UserRepositoryImpl` provides implementations of the `findById()`, `all()`, and `echoDetails()` functions defined in `UserRepository`. The three functions implemented make use of `service` to retrieve the required data residing on the server via HTTP-appropriate requests. `findById()` calls the `showUser()` function in service to send a request to the Messenger API to retrieve the details of the user with the specified user ID. The `showUser()` function requires the authorization token of the currently logged-in user as its second argument. We provide this required token via the `AppPreferences` instance by passing `preferences.accessToken` as the second argument to the function.

The `all()` function makes use of `MessengerApiService#listUsers()` to retrieve all the users that are registered on the messenger service. The `echoDetails()` function makes use of the `MessengerApiService#echoDetails()` function to get the details of the currently logged-in user.

Let's create a conversation repository to facilitate the access of data pertaining to conversations. Add a `ConversationRepository` interface to `com.example.messenger.data.remote.repository` with the following content:

```
package com.example.messenger.data.remote.repository

import com.example.messenger.data.vo.ConversationListVO
import com.example.messenger.data.vo.ConversationVO
import io.reactivex.Observable

interface ConversationRepository {
    fun findConversationById(id: Long): Observable<ConversationVO>

    fun all(): Observable<ConversationListVO>
}
```

Now create a corresponding `ConversationRepositoryImpl` class in the package, as follows:

```
package com.example.messenger.data.remote.repository

import android.content.Context
import com.example.messenger.service.MessengerApiService
import com.example.messenger.data.local.AppPreferences
import com.example.messenger.data.vo.ConversationListVO
import com.example.messenger.data.vo.ConversationVO
import io.reactivex.Observable

class ConversationRepositoryImpl(ctx: Context) : ConversationRepository {

    private val preferences: AppPreferences = AppPreferences.create(ctx)
```

```
private val service: MessengerApiService = MessengerApiService
                                            .getInstance()
```

It retrieves information pertaining to a conversation with the requested conversation ID from the Messenger API:

```
override fun findConversationById(id: Long): Observable<ConversationVO> {
    return service.showConversation(id, preferences.accessToken as String)
}
```

It retrieves all active conversations of current user from API when invoked:

```
override fun all(): Observable<ConversationListVO> {
    return service.listConversations(preferences.accessToken as String)
}
}
```

`findConversationById(Long)` retrieves the conversation thread with the corresponding ID passed to the function. The `all()` function simply retrieves all of the current user's active conversations.

Creating the login interactor

It is time to create our login interaction, which will serve as the model that the login presenter will interact with. Create a `LoginInteractor` interface in the `login` package containing the following code:

```
package com.example.messenger.ui.login

import com.example.messenger.data.local.AppPreferences
import com.example.messenger.ui.auth.AuthInteractor

interface LoginInteractor : AuthInteractor {

    interface OnDetailsRetrievalFinishedListener {
        fun onDetailsRetrievalSuccess()
        fun onDetailsRetrievalError()
    }

    fun login(username: String, password: String,
            listener: AuthInteractor.onAuthFinishedListener)

    fun retrieveDetails(preferences: AppPreferences,
            listener: OnDetailsRetrievalFinishedListener)
}
```

As you may have noticed, `LoginInteractor` extends `AuthInteractor`. This is similar to the way that `LoginView` extends `AuthView`. The `AuthInteractor` interface declares behaviors and characteristics that must be implemented by any interactor that handles authentication-related logic. Let's implement the `AuthInteractor` interface now.

Go ahead and add an `AuthInteractor` interface to the `com.exampla.messenger.auth` package:

```
package com.example.messenger.ui.auth

import com.example.messenger.data.local.AppPreferences
import com.example.messenger.data.remote.vo.UserVO

interface AuthInteractor {

    var userDetails: UserVO
    var accessToken: String
    var submittedUsername: String
    var submittedPassword: String

    interface onAuthFinishedListener {
        fun onAuthSuccess()
        fun onAuthError()
        fun onUsernameError()
        fun onPasswordError()
    }

    fun persistAccessToken(preferences: AppPreferences)

    fun persistUserDetails(preferences: AppPreferences)

}
```

Every interactor that is an `AuthInteractor` must have the following fields: `userDetails`, `accessToken`, `submittedUsername`, and `submittedPassword`. In addition, an interactor that implements `AuthInteractor` must have `persistAccessToken(AppPreferences)` and `persistUserDetails(AppPreferences)` methods. As the names of the methods suggest, they persist access tokens and user details to the application's `SharedPreferences` file. As you might have guessed, we need to create an implementation class for the `LoginInteractor`. We will call this class `LoginInteractorImpl`.

The following is the `LoginInteractorImpl` class with its implemented `login()` method. Add it to the `login` package within the `ui` package:

```
package com.example.messenger.ui.login

import com.example.messenger.data.local.AppPreferences
import com.example.messenger.data.remote.request.LoginRequestObject
import com.example.messenger.data.vo.UserVO
import com.example.messenger.service.MessengerApiService
import com.example.messenger.ui.auth.AuthInteractor
import io.reactivex.android.schedulers.AndroidSchedulers
import io.reactivex.schedulers.Schedulers


class LoginInteractorImpl : LoginInteractor {

    override lateinit var userDetails: UserVO
    override lateinit var accessToken: String
    override lateinit var submittedUsername: String
    override lateinit var submittedPassword: String

    private val service: MessengerApiService = MessengerApiService
                                        .getInstance()

    override fun login(username: String, password: String,
                    listener: AuthInteractor.onAuthFinishedListener) {
        when {
```

If an empty `username` is submitted in the login form, the `username` is invalid. The listener's `onUsernameError()` function is called when this happens:

```
username.isBlank() -> listener.onUsernameError()
```

Call the listener's `onPasswordError()` function when an empty `password` is submitted:

```
password.isBlank() -> listener.onPasswordError()
else -> {
```

Initializing model's `submittedUsername` and `submittedPassword` fields and creating appropriate `LoginRequestObject`:

```
submittedUsername = username
submittedPassword = password
val requestObject = LoginRequestObject(username, password)
```

Using `MessengerApiService` to send a login request to Messenger API.

```
service.login(requestObject)
      .subscribeOn(Schedulers.io())
        // subscribing Observable to Scheduler thread
      .observeOn(AndroidSchedulers.mainThread())
        // setting observation to be done on the main thread
      .subscribe({ res ->
        if (res.code() != 403) {
          accessToken = res.headers()["Authorization"] as String
          listener.onAuthSuccess()
        } else {
```

Branched reached when an HTTP 403 (forbidden) status code is returned by the server. This indicates that the login failed and the user is not
authorized to access the server.

```
                listener.onAuthError()
          }
        }, { error ->
          listener.onAuthError()
          error.printStackTrace()
        })
      }
    }
  }
}
```

`login()` works by first verifying that the provided `username` and `password` arguments are not blank. The `onUsernameError()` function of the `onAuthFinishedListener` is invoked when a blank username is encountered and `onPasswordError()` is invoked when a blank password is encountered. If neither the username nor password provided is blank, it then makes use of `MessengerApiService` to send a login request to the messenger API. If the login request is successful, then it sets the `accessToken` property to the access token retrieved from the `Authorization` header of the API response and then invoked the listener's `onAuthSuccess()` function. In a scenario when the login request fails, the `onAuthError()` listener function is invoked.

Having understood the login process, add the `retrieveDetails()`, `persistAccessToken()` and `persistUserDetails()` methods below to `LoginInteractorImpl`:

```
override fun retrieveDetails(preferences: AppPreferences,
            listener: LoginInteractor.OnDetailsRetrievalFinishedListener)
{
```

It retrieves details of user upon initial login:

```
service.echoDetails(preferences.accessToken as String)
        .subscribeOn(Schedulers.io())
        .observeOn(AndroidSchedulers.mainThread())
        .subscribe({ res ->
          userDetails = res
          listener.onDetailsRetrievalSuccess()},
        { error ->
          listener.onDetailsRetrievalError()
          error.printStackTrace()})
}

override fun persistAccessToken(preferences: AppPreferences) {
  preferences.storeAccessToken(accessToken)
}

override fun persistUserDetails(preferences: AppPreferences) {
  preferences.storeUserDetails(userDetails)
}
```

Make sure to read through the comments in the preceding code snippet carefully. They thoroughly explain the workings of the `LoginInteractor`. It is now time to work on the `LoginPresenter`.

Creating the login presenter

A presenter, as we saw in `Chapter 3`, *Implementing Tetris Logic and Functionality*, is the middleman between a view and a model. It is necessary to create suitable presenters for views to properly facilitate clean view-model interactions. Creating a presenter is fairly easy. We need to first create an interface that properly declares the behaviors that will be exhibited by the presenter. Create a `LoginPresenter` interface in the `login` package with the following code:

```
package com.example.messenger.ui.login

interface LoginPresenter {
  fun executeLogin(username: String, password: String)
}
```

As can be easily seen in the preceding code snippet, we want a class that acts as a
`LoginPresenter` for a `LoginView` to possess an `executeLogin(String, String)`
function. This function will be called by the view and will then interact with a model
handling the login logic for the application. We will need to create a `LoginPresenterImpl`
class that implements `LoginPresenter`:

```
package com.example.messenger.ui.login

import com.example.messenger.data.local.AppPreferences
import com.example.messenger.ui.auth.AuthInteractor


class LoginPresenterImpl(private val view: LoginView) :
        LoginPresenter, AuthInteractor.onAuthFinishedListener,
        LoginInteractor.OnDetailsRetrievalFinishedListener {

  private val interactor: LoginInteractor = LoginInteractorImpl()
  private val preferences: AppPreferences =
AppPreferences.create(view.getContext())

  override fun onPasswordError() {
    view.hideProgress()
    view.setPasswordError()
  }

  override fun onUsernameError() {
    view.hideProgress()
    view.setUsernameError()
  }

  override fun onAuthSuccess() {
    interactor.persistAccessToken(preferences)
    interactor.retrieveDetails(preferences, this)
  }

  override fun onAuthError() {
    view.showAuthError()
    view.hideProgress()
  }

  override fun onDetailsRetrievalSuccess() {
    interactor.persistUserDetails(preferences)
    view.hideProgress()
    view.navigateToHome()
  }

  override fun onDetailsRetrievalError() {
```

```
        interactor.retrieveDetails(preferences, this)
    }

    override fun executeLogin(username: String, password: String) {
        view.showProgress()
        interactor.login(username, password, this)
    }
}
```

The `LoginPresenterImpl` class implements `LoginPresenter`,
`AuthInteractor.onAuthFinishedListener`, and
`LoginInteractor.OnDetailsRetrievalFinishedListener`, and, as such, implements
all behaviors required by the interfaces. `LoginPresenterImpl` overrides seven functions in
all: `onPasswordError()`, `onUsernameError()`, `onAuthSuccess()`, `onAuthError()`,
`onDetailsRetrievalSuccess()`, `onDetailsRetrievalError()`, and
`executeLogin(String, String)`. The interaction between the `LoginPresenter` and
`LoginInteractor` can be seen within the `onAuthSuccess()` and
`executeLogin(String, String)` functions. When a user submits their login details, the
`LoginView` calls the `executeLogin(String, String)` function in `LoginPresenter`. In
turn, `LoginPresenter` uses `LoginInteractor` to handle the actual login procedure by
calling the `login(String, String)` function of `LoginInteractor`.

If the user login is successful, the `onAuthSuccess()` callback function of `LoginPresenter`
is invoked by the `LoginInteractor`. This then leads to the storing of the access token
returned by the server and the retrieval of the logged-in user's account details. When the
login request is declined by the server, `onAuthError()` is called and an informative error
message is displayed to the user.

When a user's account details are successfully retrieved by the interactor, the
`onDetailsRetrievalSuccess()` callback of `LoginPresenter` is invoked. This leads to
the storage of the account details. The progress bar shown to the user over the course of the
login is then hidden with `view.hideProgress()`, after which the user is navigated to the
home screen with `view.navigateToHome()`. If the retrieval of user details fails,
`onDetailsRetrievalError()` is called by `LoginInteractor`. The presenter then
requests another attempt at retrieving the user's account details by calling
`interactor.retrieveDetails(preferences, this)` once more.

Finishing the LoginView

If you recall, we did not finish our implementation of the `LoginView` earlier. Functions such as `navigateToSignUp()`, `navigateToHome()`, and `onClick(view: View)` were left with empty bodies. In addition, the `LoginView` did not interact in any way with the `LoginPresenter`. Let's fix that now, shall we?

Fist things first, to navigate a user to the signup screen and home screen, we need views for these screens to exist. We won't concern ourselves with implementing layouts for them now (that will be done in the following sections). We just need them to exist. Create the `signup` and `main` packages under `com.example.messenger.ui`. Create a new empty activity called `SignUpActivity` in the `signup` package and an empty activity called `MainActivity` within `main`.

Now open `LoginActivity.kt`. We need to modify the previously mentioned functions to perform their respective tasks. In addition, we need to add private properties for a `LoginPresenter` instance and an `AppPreferences` instance. These changes are made in the following code snippet:

Firstly, add the properties below to the top of the `LoginActivity` class.

```
private lateinit var progressBar: ProgressBar
private lateinit var presenter: LoginPresenter
private lateinit var preferences: AppPreferences
```

Now modify `navigateToSignUp()`, `navigateToHome()`, and `onClick(view: View)` as shown in the following snippet:

```
override fun navigateToSignUp() {
  startActivity(Intent(this, SignUpActivity::class.java))
}

override fun navigateToHome() {
  finish()
  startActivity(Intent(this, MainActivity::class.java))
}

override fun onClick(view: View) {
  if (view.id == R.id.btn_login) {
    presenter.executeLogin(etUsername.text.toString(),
                           etPassword.text.toString())
  } else if (view.id == R.id.btn_sign_up) {
    navigateToSignUp()
  }
}
```

`navigateToSignUp()` uses an explicit intent to start `SignUpActivity` when called. `navigateToHome()` operates similarly to `navigateToSignUp()`—it starts `MainActivity`. A major difference between `navigateToHome()` and `navigateToSignUp()` is that `navigateToHome()` destroys the current `LoginActivity` instance by calling `finish()` before starting the `MainActivity`.

The `onClick()` method uses the `LoginPresenter` to begin the login process in the scenario that the login button is clicked. Otherwise, if the signup button is clicked, the `SignUpActivity` is started with `navigateToSignUp()`.

Great job thus far! We have created the necessary view, presenter, and model for login-related application logic. We need to keep in mind that before we can log in a user, we need to have registered a user on the platform first. Thus, we must implement our signup logic. We will do this in the following section.

Developing the signup UI

Let's develop the signup user interface. First, we have to implement the necessary views, starting from the layout of `SignUpActivity`. We do not need much in terms of elements for our `SignUpActivity` layout. We need three input fields to take the username, password, and phone number of a user to be registered. In addition, we need a button to submit the signup form as well as a progress bar to show when the signup is in progress.

The following is our `activity_sign_up.xml` layout:

```xml
<?xml version="1.0" encoding="utf-8"?>
<android.support.constraint.ConstraintLayout
xmlns:android="http://schemas.android.com/apk/res/android"
    xmlns:tools="http://schemas.android.com/tools"
    android:layout_width="match_parent"
    android:layout_height="match_parent"
    tools:context=".ui.signup.SignUpActivity"
    android:paddingTop="@dimen/default_padding"
    android:paddingBottom="@dimen/default_padding"
    android:paddingStart="@dimen/default_padding"
    android:paddingEnd="@dimen/default_padding"
    android:orientation="vertical"
    android:gravity="center_horizontal">
    <EditText
        android:id="@+id/et_username"
        android:layout_width="match_parent"
        android:layout_height="wrap_content"
        android:hint="@string/username"
```

```
            android:inputType="text"/>
        <EditText
            android:id="@+id/et_phone"
            android:layout_width="match_parent"
            android:layout_height="wrap_content"
            android:layout_marginTop="@dimen/default_margin"
            android:hint="@string/phone_number"
            android:inputType="phone"/>
        <EditText
            android:id="@+id/et_password"
            android:layout_width="match_parent"
            android:layout_height="wrap_content"
            android:layout_marginTop="@dimen/default_margin"
            android:hint="@string/password"
            android:inputType="textPassword"/>
        <Button
            android:id="@+id/btn_sign_up"
            android:layout_width="wrap_content"
            android:layout_height="wrap_content"
            android:layout_marginTop="@dimen/default_margin"
            android:text="@string/sign_up"/>
        <ProgressBar
            android:id="@+id/progress_bar"
            android:layout_width="wrap_content"
            android:layout_height="wrap_content"
            android:layout_marginTop="@dimen/default_margin"
            android:visibility="gone"/>
</android.support.constraint.ConstraintLayout>
```

The visual translation of the XML layout written earlier is as follows:

As you can see, the layout we designed contains all the necessary elements we previously mentioned.

Creating the signup interactor

Now we will implement a signup interactor to act as a model that our yet-to-be-implemented signup presenter will communicate with. Create a `SignUpInteractor` interface within the `signup` package, as follows:

```
package com.example.messenger.ui.signup

import com.example.messenger.ui.auth.AuthInteractor


interface SignUpInteractor : AuthInteractor {

  interface OnSignUpFinishedListener {
    fun onSuccess()
    fun onUsernameError()
```

```
        fun onPasswordError()
        fun onPhoneNumberError()
        fun onError()
    }

    fun signUp(username: String, phoneNumber: String, password: String,
               listener: OnSignUpFinishedListener)

    fun getAuthorization(listener: AuthInteractor.onAuthFinishedListener)
}
```

You may have noticed that SignUpInteractor extends AuthInteractor. Similar to a LoginInteractor, a SignUpInteractor requires the use of userDetails, accessToken, submittedUsername, and submittedPassword properties. In addition, SignUpInteractor needs to be able to persist the access token of a user as well as their user details by using the persistAccessToken(AppPreferences) and persistUserDetails(AppPreferences) functions, which are declared in the AuthInteractor interface.

We created an OnSignUpFinishedListener interface within SignUpInteractor declaring callbacks that must be implemented by an OnSignUpFinishedListener. This listener will be our SignUpPresenter when we implement it.

In creating SignUpInteractorImpl, we shall start first and foremost with its property declarations and the implementation of its login() method. Create SignUpInteractorImpl which is as follows. Ensure that you add it to the same package as SignUpInteractor:

```
package com.example.messenger.ui.signup

import android.text.TextUtils
import android.util.Log
import com.example.messenger.data.local.AppPreferences
import com.example.messenger.data.remote.request.LoginRequestObject
import com.example.messenger.data.remote.request.UserRequestObject
import com.example.messenger.data.vo.UserVO
import com.example.messenger.service.MessengerApiService
import com.example.messenger.ui.auth.AuthInteractor
import io.reactivex.android.schedulers.AndroidSchedulers
import io.reactivex.schedulers.Schedulers

class SignUpInteractorImpl : SignUpInteractor {

    override lateinit var userDetails: UserVO
    override lateinit var accessToken: String
```

```
override lateinit var submittedUsername: String
override lateinit var submittedPassword: String

private val service: MessengerApiService = MessengerApiService
                                             .getInstance()

override fun signUp(username: String,
                   phoneNumber: String, password: String,
                   listener: SignUpInteractor.OnSignUpFinishedListener){
  submittedUsername = username
  submittedPassword = password
  val userRequestObject = UserRequestObject(username, password,
                                             phoneNumber)

  when {
    TextUtils.isEmpty(username) -> listener.onUsernameError()
    TextUtils.isEmpty(phoneNumber) -> listener.onPhoneNumberError()
    TextUtils.isEmpty(password) -> listener.onPasswordError()
    else -> {
```

Registering a new user to the Messenger platform with the `MessengerApiService`

```
      service.createUser(userRequestObject)
            .subscribeOn(Schedulers.io())
            .observeOn(AndroidSchedulers.mainThread())
            .subscribe({ res ->
        userDetails = res
        listener.onSuccess()
      }, { error ->
        listener.onError()
        error.printStackTrace()
      })
    }
  }
 }
}
```

Now add the `getAuthorization()`, `persistAccessToken()` and
`persistUserDetails()` methods below to `SignUpInteractorImpl`:

```
override fun getAuthorization(listener:
                AuthInteractor.onAuthFinishedListener) {
  val userRequestObject = LoginRequestObject(submittedUsername,
                                             submittedPassword)
```

Let's log the registered user in to the platform with the `MessengerApiService`:

```
service.login(userRequestObject)
        .subscribeOn(Schedulers.io())
        .observeOn(AndroidSchedulers.mainThread())
        .subscribe( { res ->
    accessToken = res.headers()["Authorization"] as String
```

Now, user has been successfully logged in. Hence, we invoke listener's `onAuthSuccess()` callback:

```
    listener.onAuthSuccess()

  }, { error ->
    listener.onAuthError()
    error.printStackTrace()
  })
}

override fun persistAccessToken(preferences: AppPreferences) {
  preferences.storeAccessToken(accessToken)
}

override fun persistUserDetails(preferences: AppPreferences) {
  preferences.storeUserDetails(userDetails)
}
```

The `SignUpInteractorImpl` class is a straightforward implementation of the `SignUpInteractor` interface. Lines 19 to 22 contain property declarations for `userDetails`, `accessToken`, `submittedUsername`, and `submittedPassword` that must be possessed by an `AuthInteractor`. `signUp(String, String, String, SignUpInteractor.OnSignUpFinishedListener)` contains the signup logic for the application. If all values submitted by a user are valid, then the user is registered on the platform with the `createUser(UserRequestObject)` function of the `MessengerApiService` that we created with Retrofit.

`getAuthorization(AuthInteractor.onAuthFinishedListener)` is called to authorize a newly registered user of the messenger platform. Make sure to peruse the comments within `SignUpInteractorImpl` for more information.

Next on our agenda is the creation of the `SignUpPresenter`.

Creating the signup presenter

As we did when creating the `LoginPresenter`, we need to create a `SignUpPresenter` interface along with a `SignUpPresenterImpl` class. The `SignUpPresenter` we are making is in no way complex. For this application, we need our signup presenter to possess a property of the `AppPreferences` type as well as a function that executes the signup process. The following is the `SignUpPresenter` interface:

```
package com.example.messenger.ui.signup

import com.example.messenger.data.local.AppPreferences

interface SignUpPresenter {
  var preferences: AppPreferences

  fun executeSignUp(username: String, phoneNumber: String, password:
String)
}
```

Now, here is the code for our `SignUpPresenter` implementation:

```
package com.example.messenger.ui.signup

import com.example.messenger.data.local.AppPreferences
import com.example.messenger.ui.auth.AuthInteractor

class SignUpPresenterImpl(private val view: SignUpView): SignUpPresenter,
                         SignUpInteractor.OnSignUpFinishedListener,
                         AuthInteractor.onAuthFinishedListener {

  private val interactor: SignUpInteractor = SignUpInteractorImpl()
  override var preferences: AppPreferences = AppPreferences
                                      .create(view.getContext())
```

The `onSuccess()` callback below is invoked when user is successfully signed up:

```
override fun onSuccess() {
   interactor.getAuthorization(this)
}
```

The callback invoked when an error occurs during user sign up:

```
override fun onError() {
  view.hideProgress()
  view.showSignUpError()
}
```

```
    override fun onUsernameError() {
      view.hideProgress()
      view.setUsernameError()
    }

    override fun onPasswordError() {
      view.hideProgress()
      view.setPasswordError()
    }

    override fun onPhoneNumberError() {
      view.hideProgress()
      view.setPhoneNumberError()
    }

    override fun executeSignUp(username: String, phoneNumber: String,
                            password: String) {
      view.showProgress()
      interactor.signUp(username, phoneNumber, password, this)
    }

    override fun onAuthSuccess() {
      interactor.persistAccessToken(preferences)
      interactor.persistUserDetails(preferences)
      view.hideProgress()
      view.navigateToHome()
    }

    override fun onAuthError() {
      view.hideProgress()
      view.showAuthError()
    }
  }
}
```

The preceding `SignUpPresenterImpl` class implements the `SignUpPresenter`, `SignUpInteractor.OnSignUpFinishedListener`, and `AuthInteractor.onAuthFinishedListener` interfaces, and, as such, provides implementations for a number of required functions. These functions are `onSuccess()`, `onError()`, `onUsernameError()`, `onPasswordError()`, `onPhoneNumberError()`, `executeSignUp(String, String, String)`, `onAuthSuccess()`, and `onAuthError()`. `SignUpPresenterImpl` takes a single argument as its primary constructor. This argument must be of the `SignUpView` type.

`executeSignUp(String, String, String)` is a function that will be invoked by a `SignUpView` to begin the user registration process. `onSuccess()` is called when a user's signup request is successful. The function immediately invokes the interactor's `getAuthorization()` function to get an access token for the newly registered user. In a scenario when a signup request fails, the `onError()` callback is invoked. This hides the progress bar being shown to the user and displays an appropriate error message.

The `onUsernameError()`, `onPasswordError()`, and `onPhoneNumberError()` methods are callbacks invoked upon the occurrence of an error in a submitted username, password, or phone number, respectively. `onAuthSuccess()` is a callback invoked when the authorization procedure is successful. On the other hand, `onAuthError()` is invoked when the authorization fails.

Creating the signup view

It is time to work on the `SignUpView`. First we need to create a `SignUpView` interface, after which we will make `SignUpActivity` implement this interface. Note that in our application, a `SignUpView` is an extension of `BaseView` and `AuthView`. The following is the `SignUpView` interface:

```
package com.example.messenger.ui.signup

import com.example.messenger.ui.auth.AuthView
import com.example.messenger.ui.base.BaseView

interface SignUpView : BaseView, AuthView {

    fun showProgress()
    fun showSignUpError()
    fun hideProgress()
    fun setUsernameError()
    fun setPhoneNumberError()
    fun setPasswordError()
    fun navigateToHome()
}
```

Now we shall modify the `SignUpActivity` class in the project to implement the `SignUpView` and make use of the `SignUpPresenter`. Add the changes in the following code snippet to `SignUpActivity`:

```
package com.example.messenger.ui.signup

import android.content.Context
```

```kotlin
import android.content.Intent
import android.support.v7.app.AppCompatActivity
import android.os.Bundle
import android.view.View
import android.widget.Button
import android.widget.EditText
import android.widget.ProgressBar
import android.widget.Toast
import com.example.messenger.R
import com.example.messenger.data.local.AppPreferences
import com.example.messenger.ui.main.MainActivity

class SignUpActivity : AppCompatActivity(), SignUpView,
View.OnClickListener {

  private lateinit var etUsername: EditText
  private lateinit var etPhoneNumber: EditText
  private lateinit var etPassword: EditText
  private lateinit var btnSignUp: Button
  private lateinit var progressBar: ProgressBar
  private lateinit var presenter: SignUpPresenter

  override fun onCreate(savedInstanceState: Bundle?) {
    super.onCreate(savedInstanceState)
    setContentView(R.layout.activity_sign_up)
    presenter = SignUpPresenterImpl(this)
    presenter.preferences = AppPreferences.create(this)
    bindViews()
  }

  override fun bindViews() {
    etUsername = findViewById(R.id.et_username)
    etPhoneNumber = findViewById(R.id.et_phone)
    etPassword = findViewById(R.id.et_password)
    btnSignUp = findViewById(R.id.btn_sign_up)
    progressBar = findViewById(R.id.progress_bar)
    btnSignUp.setOnClickListener(this)
  }

  override fun showProgress() {
    progressBar.visibility = View.VISIBLE
  }

  override fun hideProgress() {
    progressBar.visibility = View.GONE
  }

  override fun navigateToHome() {
```

```
      finish()
      startActivity(Intent(this, MainActivity::class.java))
    }

  override fun onClick(view: View) {
    if (view.id == R.id.btn_sign_up) {
      presenter.executeSignUp(etUsername.text.toString(),
                              etPhoneNumber.text.toString(),
                              etPassword.text.toString())
    }
  }
}
```

Now add the `setUsernameError()`, `setPasswordError()`, `showAuthError()`, `showSignUpError()` and `getContext()` functions shown below to `SignUpActivity`:

```
override fun setUsernameError() {
  etUsername.error = "Username field cannot be empty"
}

override fun setPhoneNumberError() {
  etPhoneNumber.error = "Phone number field cannot be empty"
}

override fun setPasswordError() {
  etPassword.error = "Password field cannot be empty"
}

override fun showAuthError() {
  Toast.makeText(this, "An authorization error occurred.
                        Please try again later.",
                Toast.LENGTH_LONG).show()
}

override fun showSignUpError() {
  Toast.makeText(this, "An unexpected error occurred.
                        Please try again later.",
                Toast.LENGTH_LONG).show()
}

override fun getContext(): Context {
  return this
}
```

Great work thus far! At this point, we are half of the way through the development of the Messenger application. You deserve a round of applause for your efforts. But we still have some work to do—especially with the main UI. We will finish up the remainder of the Messenger application in the next chapter.

Summary

In this chapter, we began the development of the Messenger Android application. In the process of doing so, we covered a vast array of topics. We got down and dirty with the Model-View-Presenter pattern and explored in detail how to create an application utilizing this modern development approach.

Further into the chapter, we learned about reactive programming and made extensive use of RxJava and RxAndroid. We learned how to communicate with a remote server using OkHttp and Retrofit, after which we took things one step further and implemented a fully functional Retrofit service to communicate with the Messenger API we developed in Chapter 4, *Designing and Implementing the Messenger Backend with Spring Boot 2.0*.

In the next chapter, we will finish our work with the Messenger application.

6
Building the Messenger Android App – Part II

In the previous chapter, we went full steam ahead with the development of the Messenger Android application. By doing so, we examined both Kotlin and Android application development in depth. We explored the **Model-View-Presenter** (**MVP**) pattern and how to use it to build powerful and fully functional Android applications. In addition to this, we covered the basics of Reactive programming and learned how to use RxJava and RxAndroid in our applications. We also learned about some of the available means by which we can communicate with a remote server. We learned about OkHttp and Retrofit, and then went one step further by implementing a fully functional Retrofit service to facilitate communication with the messenger API that we made in `Chapter 4`, *Designing and Implementing the Messenger Backend with Spring Boot 2.0*. Putting all this knowledge pertaining to Android and Kotlin together, we created both a login and signup user interface for the Messenger app.

In this chapter, we will finish the development of the Messenger app. In the process of doing so, we will cover the following topics:

- Working with application settings
- Working with ChatKit
- Android application testing
- Performing background tasks

Let's continue the development of our Messenger app by implementing the Main UI.

Creating the Main UI

Similar to what we have done in implementing the Login UI and SignUp UI, we will create a model, view, and presenter for the Main UI. We are not going to focus as much on explanations as we did in the process of implementing the previous two UI views. Instead, only new concepts will be explained.

Without further ado, let's create a `MainView`.

Creating the MainView

Before we proceed with creating the main view, it is imperative that we have a clear understanding of the user interface that we want to implement. A good place to start is to clearly write out sentences that describe how we want the `MainView` to function. Let's go ahead and do that:

- The main view should display the active conversations of the currently logged-in user upon launch
- The main view should allow a logged-in user to create a new conversation
- The main view should be able to show the contacts of a currently logged-in user (in the case of this application, this is a list of all the registered users on the Messenger platform)
- A user must be able to access the settings screen directly from the `MainView`
- A user should be able to log out of the application directly from the `MainView`

All right, great! We have our list of brief statements describing what the `MainView` can do. With this list, it is possible to get on with creating the `MainView` (in terms of programming, that is). We are not going to do this yet. Let's create a few visual sketches of `MainView` to give us a clearer idea of how it will look:

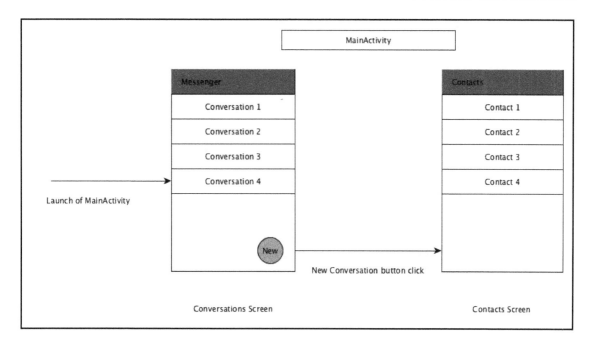

As can be seen from the preceding diagram, MainActivity can render two completely separate views to a user. The first view is the **Conversations Screen** and the second the **Contacts Screen**. A perfect way to implement this is to employ two use fragments within the MainActivity. In this case, we will require two distinct fragments. These are the conversations fragment and the contacts fragment.

Now that we have a clear idea of what the MainView is going to contain, we need to implement a proper interface to declare the behaviors of the MainView. The following is the MainView interface:

```
package com.example.messenger.ui.main
import com.example.messenger.ui.base.BaseView

interface MainView : BaseView {
  fun showConversationsLoadError()
  fun showContactsLoadError()
  fun showConversationsScreen()
  fun showContactsScreen()
  fun getContactsFragment(): MainActivity.ContactsFragment
  fun getConversationsFragment(): MainActivity.ConversationsFragment
  fun showNoConversations()
  fun navigateToLogin()
  fun navigateToSettings()
```

```
}
```

Great work! We will save the implementation of the `MainView` by `MainActivity` for later. For now, we will work on the `MainInteractor`.

Creating the MainInteractor

We want the user to be able to view other users (contacts) on the Messenger platform and view their active conversations on the main screen. In addition to this, we want a user to be able to log out of the platform directly from the main screen. As a result of these requirements, `MainInteractor` must be capable of loading contacts, loading conversations, and logging the user out of the platform. The following is the `MainInteractor` interface. Ensure to put it and all other `Main_` files in the `com.example.messenger.ui.main` package:

```
package com.example.messenger.ui.main

import com.example.messenger.data.vo.ConversationListVO
import com.example.messenger.data.vo.UserListVO

interface MainInteractor {

  interface OnConversationsLoadFinishedListener {
    fun onConversationsLoadSuccess(
    conversationsListVo: ConversationListVO)

      fun onConversationsLoadError()
  }

  interface OnContactsLoadFinishedListener {
    fun onContactsLoadSuccess(userListVO: UserListVO)
    fun onContactsLoadError()
  }

  interface OnLogoutFinishedListener {
    fun onLogoutSuccess()
  }

  fun loadContacts(
  listener: MainInteractor.OnContactsLoadFinishedListener)

  fun loadConversations(
  listener: MainInteractor.OnConversationsLoadFinishedListener)
```

```
    fun logout(listener: MainInteractor.OnLogoutFinishedListener)
  }
```

We added the `OnConversationsLoadFinishedListener`, `OnContactsLoadFinishedListener`, and `OnLogoutFinishedListener` interfaces to the `MainInteractor` interface. These are all interfaces that will be implemented by a `MainPresenter`. These callbacks are necessary for the presenter to perform appropriate actions regardless of the success or failure of a conversation load, contact load, or user logout process.

The `MainInteractorImpl` class with an implemented `loadContacts()` method is given below:

```
package com.example.messenger.ui.main

import android.content.Context
import android.util.Log
import com.example.messenger.data.local.AppPreferences
import com.example.messenger.data.remote.repository.ConversationRepository
import
com.example.messenger.data.remote.repository.ConversationRepositoryImpl
import com.example.messenger.data.remote.repository.UserRepository
import com.example.messenger.data.remote.repository.UserRepositoryImpl
import io.reactivex.android.schedulers.AndroidSchedulers
import io.reactivex.schedulers.Schedulers

class MainInteractorImpl(val context: Context) : MainInteractor {

  private val userRepository: UserRepository =
  UserRepositoryImpl(context)
  private val conversationRepository: ConversationRepository =
  ConversationRepositoryImpl(context)
  override fun loadContacts(listener:
  MainInteractor.OnContactsLoadFinishedListener) {
```

Let's load all users registered on the Messenger API platform. These users are contacts that can be communicated with by the currently logged in user:

```
    userRepository.all()
    .subscribeOn(Schedulers.io())
    .observeOn(AndroidSchedulers.mainThread())
    .subscribe({ res ->
```

Now, the contacts were loaded successfully. `onContactsLoadSuccess()` is called with the API response data passed as an argument:

```
listener.onContactsLoadSuccess(res) },
{ error ->
```

If the contact load failed, hence, `onContactsLoadError()` is called:

```
        listener.onContactsLoadError()
        error.printStackTrace()})
    }
  }
```

`loadContacts()` makes use of `UserRepository` to load a list of all available users on the messenger platform. If the users were successfully retrieved, the listener's `onContactsLoadSuccess()` is invoked with the list of the users loaded passed as an argument. Otherwise, `onContactsLoadError()` is invoked and the error is printed to the standard system output .

We are not done with `MainInteractorImpl` yet. We must still add functions for `loadConversations()` and `logout()`. These two required functions are given in the following code snippet. Add them to `MainInteractorImpl`.

```
override fun loadConversations(
listener: MainInteractor.OnConversationsLoadFinishedListener) {
```

It retrieves all conversations of the currently logged in user using conversational repository instance:

```
    conversationRepository.all()
    .subscribeOn(Schedulers.io())
    .observeOn(AndroidSchedulers.mainThread())
    .subscribe({ res -> listener.onConversationsLoadSuccess(res) },
    { error ->
      listener.onConversationsLoadError()
    error.printStackTrace()})
  }

  override fun logout(
  listener: MainInteractor.OnLogoutFinishedListener) {
```

When login out clear user data from shared preferences file and invokes listener's `onLogoutSuccess()` callback:

```
    val preferences: AppPreferences = AppPreferences.create(context)
    preferences.clear()
```

```
    listener.onLogoutSuccess()
  }
```

`loadConversations()` works similarly to `loadContacts()`. The difference being that `ConversationRepository` is being used to retrieve active conversations that the user currently has instead of a list of contacts. `logout()` simply clears the preferences file used by the application to remove the currently logged in user's data, after which the `onLogoutSuccess()` method of the provided `OnLogoutFinishedListener` invoked.

That's all for the `MainInteractorImpl` class. Next on our agenda is the implementation of the `MainPresenter`.

Creating the MainPresenter

As always, the first thing we must do is create a presenter interface that defines functions to be implemented by a presenter implementation class. The following is the `MainPresenter` interface:

```
package com.example.messenger.ui.main

interface MainPresenter {
  fun loadConversations()
  fun loadContacts()
  fun executeLogout()
}
```

The `loadConversations()`, `loadContacts()`, and `executeLogout()` functions will be invoked by the `MainView` and must be implemented by the `MainPresenterImpl` class. Our `MainPresenterImpl` class with its defined properties, and `onConversationsLoadSuccess()` and `onConversationsLoadError()` methods is given as follows:

```
package com.iyanuadelekan.messenger.ui.main

import com.iyanuadelekan.messenger.data.vo.ConversationListVO
import com.iyanuadelekan.messenger.data.vo.UserListVO

class MainPresenterImpl(val view: MainView) : MainPresenter,
        MainInteractor.OnConversationsLoadFinishedListener,
        MainInteractor.OnContactsLoadFinishedListener,
        MainInteractor.OnLogoutFinishedListener {

  private val interactor: MainInteractor = MainInteractorImpl
                                      (view.getContext())
```

```
override fun onConversationsLoadSuccess(conversationsListVo:
                                        ConversationListVO) {
```

Let's check if currently logged in user has active conversations:

```
if (!conversationsListVo.conversations.isEmpty()) {
    val conversationsFragment = view.getConversationsFragment()
    val conversations = conversationsFragment.conversations
    val adapter = conversationsFragment.conversationsAdapter

    conversations.clear()
    adapter.notifyDataSetChanged()
```

After retrieving conversations from API, we add each conversation to `ConversationFragment`'s conversations list and conversations adapter is notified after every item addition:

```
    conversationsListVo.conversations.forEach { contact ->
        conversations.add(contact)
        adapter.notifyItemInserted(conversations.size - 1)
    }
} else {
    view.showNoConversations()
}
}

override fun onConversationsLoadError() {
    view.showConversationsLoadError()
}
}
```

In addition, add the `onContactsLoadSuccess()`, `onContactsLoadError()`, `onLogoutSuccess()`, `loadConversations()`, `loadContacts()` and `executeLogout()` functions given below to `MainPresenterImpl`:

```
override fun onContactsLoadSuccess(userListVO: UserListVO) {
    val contactsFragment = view.getContactsFragment()
    val contacts = contactsFragment.contacts
    val adapter = contactsFragment.contactsAdapter
```

Let's clear previously loaded contacts in contacts list and notify adapter pf data set change:

```
contacts.clear()
adapter.notifyDataSetChanged()
```

Now, let's add each contact retrieved from API to `ContactsFragment`'s contacts list and contacts adapter is notified after every item addition:

```
    userListVO.users.forEach { contact ->
      contacts.add(contact)
      contactsFragment.contactsAdapter.notifyItemInserted(contacts.size-1)
    }
  }

  override fun onContactsLoadError() {
    view.showContactsLoadError()
  }

  override fun onLogoutSuccess() {
    view.navigateToLogin()
  }

  override fun loadConversations() {
    interactor.loadConversations(this)
  }

  override fun loadContacts() {
    interactor.loadContacts(this)
  }

  override fun executeLogout() {
    interactor.logout(this)
  }
```

We have successfully created our `MainInteractor` and `MainPresenter`. At this point, it is time to finish up our work on the `MainView` and its layouts.

Wrapping up the MainView

First and foremost, we must work on the `activity_main.xml` layout file. Modify the file to contain the following code:

```
    <?xml version="1.0" encoding="utf-8"?>
    <android.support.design.widget.CoordinatorLayout
      xmlns:android="http://schemas.android.com/apk/res/android"
      xmlns:tools="http://schemas.android.com/tools"
      android:layout_width="match_parent"
      android:layout_height="match_parent"
      tools:context=".ui.main.MainActivity">
      <LinearLayout
```

```
    android:id="@+id/ll_container"
    android:layout_width="match_parent"
    android:layout_height="match_parent"
    android:orientation="vertical"/>
</android.support.design.widget.CoordinatorLayout>
```

Within the root view of the layout file, we have a single LinearLayout. This ViewGroup will act as a container for the conversations and contacts fragments. Speaking of conversations and contacts fragments, we must create appropriate layouts for them as well. Create a fragment_conversations.xml layout file in the project's layout resource directory with the following content:

```
<?xml version="1.0" encoding="utf-8"?>
<android.support.design.widget.CoordinatorLayout
xmlns:android="http://schemas.android.com/apk/res/android"
    android:layout_width="match_parent"
    android:layout_height="match_parent"
xmlns:app="http://schemas.android.com/apk/res-auto">
<android.support.v7.widget.RecyclerView
    android:id="@+id/rv_conversations"
    android:layout_width="match_parent"
android:layout_height="match_parent"/>
<android.support.design.widget.FloatingActionButton
    android:id="@+id/fab_contacts"
    android:layout_width="wrap_content"
    android:layout_height="wrap_content"
    android:layout_margin="@dimen/default_margin"
    android:src="@android:drawable/ic_menu_edit"
    app:layout_anchor="@id/rv_conversations"
    app:layout_anchorGravity="bottom|right|end"/>
</android.support.design.widget.CoordinatorLayout>
```

We made use of two child views within the CoordinatorLayout root view. The first is a RecyclerView and the second is a FloatingActionButton. A RecyclerView is an Android widget that is used as a container for displaying large sets of data that can be scrolled through efficiently by maintaining a limited number of views. We are able to make use of the RecyclerView widget because we added its dependency to our project's module-level build.gradle script, as follows:

```
implementation 'com.android.support:recyclerview-v7:26.1.0'
```

As we are making use of the RecyclerView widgets, we need to create appropriate view holder layouts for each RecyclerView widget. Create a vh_contacts.xml file and a vh_conversations.xml file within the layouts resource directory.

The following is the vh_contacts.xml layout:

```xml
<?xml version="1.0" encoding="utf-8"?>
<LinearLayout xmlns:android="http://schemas.android.com/apk/res/android"
  android:orientation="vertical" android:layout_width="match_parent"
  android:id="@+id/ll_container"
  android:layout_height="wrap_content">
  <LinearLayout
    android:layout_width="match_parent"
    android:layout_height="wrap_content"
    android:orientation="vertical"
    android:padding="@dimen/default_padding">
    <LinearLayout
      android:layout_width="match_parent"
      android:layout_height="wrap_content"
      android:orientation="horizontal">
      <TextView
        android:id="@+id/tv_username"
        android:layout_width="wrap_content"
        android:layout_height="wrap_content"
        android:textSize="18sp"
        android:textStyle="bold"/>
      <LinearLayout
        android:layout_width="0dp"
        android:layout_height="wrap_content"
        android:layout_weight="1"
        android:gravity="end">
        <TextView
          android:id="@+id/tv_phone"
          android:layout_width="wrap_content"
          android:layout_height="wrap_content"
          android:layout_marginLeft="@dimen/default_margin"
          android:layout_marginStart="@dimen/default_margin"/>
      </LinearLayout>
    </LinearLayout>
    <TextView
      android:id="@+id/tv_status"
      android:layout_width="wrap_content"
      android:layout_height="wrap_content"/>
  </LinearLayout>
  <View
    android:layout_width="match_parent"
    android:layout_height="1dp"
    android:background="#e8e8e8"/>
</LinearLayout>
```

The `vh_conversations.xml` layout should contain the following code:

```xml
<?xml version="1.0" encoding="utf-8"?>
<LinearLayout xmlns:android="http://schemas.android.com/apk/res/android"
  android:orientation="vertical" android:layout_width="match_parent"
  android:id="@+id/ll_container"
  android:layout_height="wrap_content">
  <LinearLayout
    android:layout_width="match_parent"
    android:layout_height="wrap_content"
    android:orientation="vertical"
    android:padding="@dimen/default_padding">
    <TextView
        android:id="@+id/tv_username"
        android:layout_width="wrap_content"
        android:layout_height="wrap_content"
        android:textStyle="bold"
        android:textSize="18sp"/>
    <TextView
        android:id="@+id/tv_preview"
        android:layout_width="wrap_content"
        android:layout_height="wrap_content"/>
  </LinearLayout>
  <View
    android:layout_width="match_parent"
    android:layout_height="1dp"
    android:background="#e8e8e8"/>
</LinearLayout>
```

As written in the Android developers reference, *Floating action buttons are used for a special type of promoted action. They are distinguished by a circled icon floating above the UI and have special motion behaviors related to morphing, launching, and the transferring anchor point.* We can make use of the `FloatingActionButton` widget because we added the Android support design library dependency to the project's `build.gradle` script:

```
implementation 'com.android.support:design:26.1.0'
```

Create a `fragment_contacts.xml` layout file within the layout resource directory containing the following XML:

```xml
<?xml version="1.0" encoding="utf-8"?>
<LinearLayout xmlns:android="http://schemas.android.com/apk/res/android"
  android:orientation="vertical" android:layout_width="match_parent"
  android:layout_height="match_parent">
  <android.support.v7.widget.RecyclerView
    android:id="@+id/rv_contacts"
    android:layout_width="match_parent"
```

```
                android:layout_height="match_parent"/>
        </LinearLayout>
```

Now it is time to finish up the `MainActivity` class. There is a lot we need to get done to complete `MainActivity`. First and foremost we must declare the necessary class properties. Next, we need to provide implementations for the following methods: `bindViews()`, `showConversationsLoadError()`, `showContactsLoadError()`, `showConversationsScreen()`, `showContactsScreen()`, `getContext()`, `getContactsFragment()`, `getConversationsFragment()`, `navigateToLogin()` and `navigateToSettings()`. Finally, we will create `ConversationsFragment` and `ContactsFragment` classes.

That is a lot to get done. We will start first and foremost with the addition of `ConversationsFragment` and `ContactsFragment` to `MainActivity`. `ConversationsFragment` is given below. Add it within `MainActivity`.

```
//ConversationsFragment class extending the Fragment class
  class ConversationsFragment : Fragment(), View.OnClickListener {

    private lateinit var activity: MainActivity
    private lateinit var rvConversations: RecyclerView
    private lateinit var fabContacts: FloatingActionButton
    var conversations: ArrayList<ConversationVO> = ArrayList()
    lateinit var conversationsAdapter: ConversationsAdapter
```

The following method is called, when user interface of `ConversationsFragment` is being drawn for the first time:

```
    override fun onCreateView(inflater: LayoutInflater, container:
    ViewGroup, savedInstanceState: Bundle?): View? {
      // fragment layout inflation
      val baseLayout =
      inflater.inflate(R.layout.fragment_conversations,
      container, false)

      // Layout view bindings
      rvConversations = baseLayout.findViewById(R.id.rv_conversations)
      fabContacts = baseLayout.findViewById(R.id.fab_contacts)

      conversationsAdapter   = ConversationsAdapter(
      getActivity(), conversations)

      // Setting the adapter of conversations recycler view to
      // created conversations adapter
      rvConversations.adapter = conversationsAdapter
```

Setting the layout manager of conversations recycler and let's see how to view a linear layout manager:

```
    rvConversations.layoutManager =
    LinearLayoutManager(getActivity().baseContext)
    fabContacts.setOnClickListener(this)
    return baseLayout
}

override fun onClick(view: View) {
    if (view.id == R.id.fab_contacts) {
        this.activity.showContactsScreen()
    }
}

fun setActivity(activity: MainActivity) {
    this.activity = activity
}
}
```

`ConversationsFragment` possesses a `RecyclerView` layout element. Recycler views need adapters to provide a binding from a data set to views that are displayed within the `RecyclerView`. Simply put, a `RecyclerView` make use of an `Adapter` to provide data for the views it renders to the display. Add `ConversationsAdapter` below as a nested class (an inner class) of `ConversationsFragment`:

```
class ConversationsAdapter(private val context:
Context, private val dataSet: List<ConversationVO>) :
RecyclerView.Adapter<ConversationsAdapter.ViewHolder>(),
ChatView.ChatAdapter {

    val preferences: AppPreferences =
    AppPreferences.create(context)

    override fun onBindViewHolder(holder: ViewHolder, position:
    Int) {
        val item = dataSet[position] // get item at current position
        val itemLayout = holder.itemLayout // bind view holder layout
        // to local variable

        itemLayout.findViewById<TextView>(R.id.tv_username).text =
        item.secondPartyUsername
        itemLayout.findViewById<TextView>(R.id.tv_preview).text =
        item.messages[item.messages.size - 1].body
```

Now, let's set `View.OnClickListener` of `itemLayout`:

```
itemLayout.setOnClickListener {
  val message = item.messages[0]
  val recipientId: Long

  recipientId = if (message.senderId ==
  preferences.userDetails.id) {
    message.recipientId
  } else {
    message.senderId
  }

  navigateToChat(item.secondPartyUsername,
  recipientId, item.conversationId)
  }
}

override fun onCreateViewHolder(parent: ViewGroup,
viewType: Int): ViewHolder {
```

Now, let's create the `ViewHolder` layout:

```
val itemLayout = LayoutInflater.from(parent.context)
.inflate(R.layout.vh_conversations, null, false)
.findViewById<LinearLayout>(R.id.ll_container)

return ViewHolder(itemLayout)
}

override fun getItemCount(): Int {
  return dataSet.size
}

override fun navigateToChat(recipientName: String,
recipientId: Long, conversationId: Long?) {
  val intent = Intent(context, ChatActivity::class.java)
  intent.putExtra("CONVERSATION_ID", conversationId)
  intent.putExtra("RECIPIENT_ID", recipientId)
  intent.putExtra("RECIPIENT_NAME", recipientName)

  context.startActivity(intent)
}

class ViewHolder(val itemLayout: LinearLayout) :
RecyclerView.ViewHolder(itemLayout)
}
```

When creating a recycler view `Adapter`, there are some important methods that you must provide custom implementations for. These methods are: `onCreateViewHolder()`, `onBindViewHolder()`, and `getItemCount()`. `onCreateViewHolder()` is invoked when the recycler view needs a new view holder instance. `onBindViewHolder()` is called by the recycler view in order to display data in the data set at a specified position. `getItemCount()` is called to get the number of items in the data set. A `ViewHolder` describes an item view in use as well as metadata about its place in a `RecyclerView`.

 An inner class is a class nested in another.

Having understood what is going on in `ConversationsFragment`, let us proceed by implementing `ContactsFragment`. First and add the following `ContactsFragment` class to `MainActivity`:

```
class ContactsFragment : Fragment() {

    private lateinit var activity: MainActivity
    private lateinit var rvContacts: RecyclerView
    var contacts: ArrayList<UserVO> = ArrayList()
    lateinit var contactsAdapter: ContactsAdapter

    override fun onCreateView(inflater: LayoutInflater,
    container: ViewGroup, savedInstanceState: Bundle?): View? {
      val baseLayout = inflater.inflate(R.layout.fragment_contacts,
      container, false)
      rvContacts = baseLayout.findViewById(R.id.rv_contacts)
      contactsAdapter = ContactsAdapter(getActivity(), contacts)

      rvContacts.adapter = contactsAdapter
      rvContacts.layoutManager =
      LinearLayoutManager(getActivity().baseContext)

      return baseLayout
    }

    fun setActivity(activity: MainActivity) {
      this.activity = activity
    }
}
```

As you most likely noticed, similar to `ConversationsFragment`, `ContactsFragment` makes use of a `RecyclerView` to render contact view elements to an application's user. The corresponding adapter class for this `RecyclerView` is `ContactsAdapter`. It is given in the following code snippet. Add it as an inner class of `ContactsFragment`:

```
class ContactsAdapter(private val context: Context,
                      private val dataSet: List<UserVO>) :
                      RecyclerView.Adapter<ContactsAdapter.ViewHolder>(),
                      ChatView.ChatAdapter {

  override fun onCreateViewHolder(parent: ViewGroup,
                                  viewType: Int): ViewHolder {
    val itemLayout = LayoutInflater.from(parent.context)
                     .inflate(R.layout.vh_contacts, parent, false)
    val llContainer = itemLayout.findViewById<LinearLayout>
                      (R.id.ll_container)

    return ViewHolder(llContainer)
  }

  override fun onBindViewHolder(holder: ViewHolder, position: Int) {
    val item = dataSet[position]
    val itemLayout = holder.itemLayout

    itemLayout.findViewById<TextView>(R.id.tv_username).text =
item.username
    itemLayout.findViewById<TextView>(R.id.tv_phone).text =
item.phoneNumber
    itemLayout.findViewById<TextView>(R.id.tv_status).text = item.status

    itemLayout.setOnClickListener {
      navigateToChat(item.username, item.id)
    }
  }

  override fun getItemCount(): Int {
    return dataSet.size
  }

  override fun navigateToChat(recipientName: String,
                              recipientId: Long, conversationId: Long?) {
    val intent = Intent(context, ChatActivity::class.java)
    intent.putExtra("RECIPIENT_ID", recipientId)
    intent.putExtra("RECIPIENT_NAME", recipientName)

    context.startActivity(intent)
  }
```

```
class ViewHolder(val itemLayout: LinearLayout) :
  RecyclerView.ViewHolder(itemLayout)
}
```

So far so good. Having created the necessary fragments, we can get to work on the properties and methods of `MainActivity`. Add the property definitions below to the top of the `MainActivity` class:

```
private lateinit var llContainer: LinearLayout
private lateinit var presenter: MainPresenter

// Creation of fragment instances
private val contactsFragment = ContactsFragment()
private val conversationsFragment = ConversationsFragment()
```

Next, modify `onCreate()` to reflect the following changes:

```
override fun onCreate(savedInstanceState: Bundle?) {
  super.onCreate(savedInstanceState)
  setContentView(R.layout.activity_main)
  presenter = MainPresenterImpl(this)

  conversationsFragment.setActivity(this)
  contactsFragment.setActivity(this)

  bindViews()
  showConversationsScreen()
}
```

Now add the `bindViews()`, `showConversationsLoadError()`, `showContactsLoadError()`, and `showConversationsScreen()` and `showContactsScreen()` methods below to `MainActivity`:

```
override fun bindViews() {
  llContainer = findViewById(R.id.ll_container)
}

override fun onCreateOptionsMenu(menu: Menu?): Boolean {
  menuInflater.inflate(R.menu.main, menu)
  return super.onCreateOptionsMenu(menu)
}

override fun showConversationsLoadError() {
  Toast.makeText(this, "Unable to load conversations.
  Try again later.",
  Toast.LENGTH_LONG).show()
}
```

```
override fun showContactsLoadError() {
  Toast.makeText(this, "Unable to load contacts. Try again later.",
  Toast.LENGTH_LONG).show()
}
```

Let's begin a new fragment transaction and replace any fragment present in activity's fragment container with a `ConversationsFragment`:

```
override fun showConversationsScreen() {
  val fragmentTransaction = fragmentManager.beginTransaction()
  fragmentTransaction.replace(R.id.ll_container,  conversationsFragment)
  fragmentTransaction.commit()

  // Begin conversation loading process
  presenter.loadConversations()

  supportActionBar?.title = "Messenger"
  supportActionBar?.setDisplayHomeAsUpEnabled(false)
}

override fun showContactsScreen() {
  val fragmentTransaction = fragmentManager.beginTransaction()
  fragmentTransaction.replace(R.id.ll_container, contactsFragment)
  fragmentTransaction.commit()
  presenter.loadContacts()

  supportActionBar?.title = "Contacts"
  supportActionBar?.setDisplayHomeAsUpEnabled(true)
}
```

Finally, add the `showNoConversations()`, `onOptionsItemSelected()`, `getContext()`, `getContactsFragment()`, `getConversationsFragment()`, `navigateToLogin()` and `navigateToSettings()` functions below to `MainActivity`:

```
override fun showNoConversations() {
  Toast.makeText(this, "You have no active conversations.",
  Toast.LENGTH_LONG).show()
}

override fun onOptionsItemSelected(item: MenuItem?): Boolean {
  when (item?.itemId) {
    android.R.id.home -> showConversationsScreen()
    R.id.action_settings -> navigateToSettings()
    R.id.action_logout -> presenter.executeLogout()
  }

  return super.onOptionsItemSelected(item)
```

```
  }

  override fun getContext(): Context {
    return this
  }

  override fun getContactsFragment(): ContactsFragment {
    return contactsFragment
  }

  override fun getConversationsFragment(): ConversationsFragment {
    return conversationsFragment
  }

  override fun navigateToLogin() {
    startActivity(Intent(this, LoginActivity::class.java))
    finish()
  }

  override fun navigateToSettings() {
    startActivity(Intent(this, SettingsActivity::class.java))
  }
```

Comments were placed in some areas within the preceding code snippets have been heavily commented to give you more understanding of what was done. Ensure you go through the comments carefully to fully grasp what we have done.

Creating the MainActivity menu

In the `onCreateOptionsMenu(Menu)` function of `MainActivity`, we inflated a menu that we have not yet implemented. Add a `main.xml` file in the `menu` package under the application resource directory. `main.xml` should contain the following content:

```xml
<?xml version="1.0" encoding="utf-8"?>
<menu xmlns:android="http://schemas.android.com/apk/res/android"
xmlns:app="http://schemas.android.com/apk/res-auto">
<item
  android:id="@+id/action_settings"
  android:orderInCategory="100"
  android:title="@string/action_settings"
  app:showAsAction="never" />
<item
  android:id="@+id/action_logout"
  android:orderInCategory="100"
  android:title="@string/action_logout"
```

```
            app:showAsAction="never" />
        </menu>
```

Fantastic work! We are one step closer to finishing this project. It is now time for us to work on the chat user The `showConversationLoadError()` and `showMessageSendError()` are functions that, interface (where the actual chatting happens).

Creating the Chat UI

The chat UI we are about to create must display the message thread of an active conversation as well as enable a user to send a new message to the individual they are chatting with. We will start this section by creating the view layout that will be rendered to the user.

Creating the chat layout

We will make use of an open source library called ChatKit to create the chat view's layout. ChatKit is an Android library that provides flexible components for chat user interface implementation in Android projects as well as utilities for chat-user-interface data management and customization.

We added ChatKit to the Messenger project with the following line of code in the `build.gradle` script:

```
    implementation 'com.github.stfalcon:chatkit:0.2.2'
```

As mentioned earlier, ChatKit provides a number of useful user interface widgets for creating a chat UI. Two of these widgets are `MessagesList` and `MessageInput`. The `MessagesList` is a widget for the display and management of messages in conversation threads. `MessageInput` is a widget for entering text messages. In addition to supporting several styling options, `MessageInput` supports simple input validation processes.

Let's see how we can use `MessagesList` and `MessageInput` in a layout file. Create a new `chat` package within `com.example.messenger.ui` and add a new empty activity named `ChatActivity` to it. Open the `ChatActivity` activities layout file (`activity_chat.xml`) and add the following XML to it:

```
    <?xml version="1.0" encoding="utf-8"?>
    <RelativeLayout xmlns:android="http://schemas.android.com/apk/res/android"
      xmlns:app="http://schemas.android.com/apk/res-auto"
```

```
    xmlns:tools="http://schemas.android.com/tools"
    android:layout_width="match_parent"
    android:layout_height="match_parent"
    tools:context="com.example.messenger.ui.chat.ChatActivity">
    <com.stfalcon.chatkit.messages.MessagesList
      android:id="@+id/messages_list"
      android:layout_width="match_parent"
      android:layout_height="match_parent"
      android:layout_above="@+id/message_input"/>
    <com.stfalcon.chatkit.messages.MessageInput
      android:id="@+id/message_input"
      android:layout_width="match_parent"
      android:layout_height="wrap_content"
      android:layout_alignParentBottom="true"
      app:inputHint="@string/hint_enter_a_message" />
</RelativeLayout>
```

As you can see in the preceding XML, we made use of ChatKit's `MessagesList` and `MessageInput` UI widgets just as we would any other Android widgets. Both `MessagesList` and `MessageInput` are located within the `com.stfalcon.chatkit.messages` package. Open the layout design window to see how the layout looks visually.

Next on our agenda is the creation of a `ChatView` class:

```
package com.example.messenger.ui.chat

import com.example.messenger.ui.base.BaseView
import com.example.messenger.utils.message.Message
import com.stfalcon.chatkit.messages.MessagesListAdapter


interface ChatView : BaseView {

  interface ChatAdapter {
    fun navigateToChat(recipientName: String, recipientId: Long,
    conversationId: Long? = null)
  }

  fun showConversationLoadError()

  fun showMessageSendError()

  fun getMessageListAdapter(): MessagesListAdapter<Message>
}
```

Within `ChatView`, we defined a `ChatAdapter` interface declaring a single `navigateToChat(String, Long, Long)` function. This interface should be implemented by adapters that are capable of directing a user to the `ChatView`. Both the `ConversationsAdapter` and `ContactsAdapter` that we earlier created implement this interface.

The `showConversationLoadError()` and `showMessageSendError()` are functions that, when implemented, must display appropriate error messages when the loading of a conversation and the loading of a message fail, respectively.

ChatKit's `MessagesList` UI widget must possess a `MessagesListAdapter` for the management of its messages dataset. `getMessageListAdapter()` is a function that, when implemented by a `ChatView`, will return the `MessagesListAdapter` of the UI's `MessagesList`.

Preparing chat UI models

To be able to add messages to the `MessagesListAdapter` of a `MessageList`, we must implement ChatKit's `IMessage` interface in an appropriate Model. We will implement this model here. Create a `com.example.messenger.utils.message` package and add the following `Message` class within it:

```
package com.example.messenger.utils.message

import com.stfalcon.chatkit.commons.models.IMessage
import com.stfalcon.chatkit.commons.models.IUser
import java.util.*

data class Message(private val authorId: Long, private val body: String,
private val createdAt: Date) : IMessage {

  override fun getId(): String {
    return authorId.toString()
  }

  override fun getCreatedAt(): Date {
    return createdAt
  }

  override fun getUser(): IUser {
    return Author(authorId, "")
  }
```

```
override fun getText(): String {
  return body
}

}
```

In addition to this, we need to create an `Author` class that implements ChatKit's `IUser` interface. The implementation of this class is as follows:

```
package com.example.messenger.utils.message

import com.stfalcon.chatkit.commons.models.IUser


data class Author(val id: Long, val username: String) : IUser {

  override fun getAvatar(): String? {
    return null
  }

  override fun getName(): String {
    return username
  }

  override fun getId(): String {
    return id.toString()
  }

}
```

The `Author` class models the user details of a message author, such as the name of the author, their ID, and an avatar (if they have one).

We have done enough with views and layouts for now. Let's go ahead and implement a `ChatInteractor` and `ChatPresenter`.

Creating the ChatInteractor and ChatPresenter

By now, we already understand what presenters and interactors are meant to do, so let's go straight to creating code. The following is the `ChatInteractor` interface. This and all other `Chat_` files belong to the `com.example.messenger.ui.chat` package:

```
package com.example.messenger.ui.chat

import com.example.messenger.data.vo.ConversationVO
```

```
interface ChatInteractor {

  interface OnMessageSendFinishedListener {
    fun onSendSuccess()

    fun onSendError()
  }

  interface onMessageLoadFinishedListener {
    fun onLoadSuccess(conversationVO: ConversationVO)
    fun onLoadError()
  }

  fun sendMessage(recipientId: Long, message: String, listener:
  OnMessageSendFinishedListener)

  fun loadMessages(conversationId: Long, listener:
  onMessageLoadFinishedListener)
}
```

The following is a corresponding ChatInteractorImpl class for
the ChatInteractor interface:

```
package com.example.messenger.ui.chat

import android.content.Context
import com.example.messenger.data.local.AppPreferences
import com.example.messenger.data.remote.repository.ConversationRepository
import
com.example.messenger.data.remote.repository.ConversationRepositoryImpl
import com.example.messenger.data.remote.request.MessageRequestObject
import com.example.messenger.service.MessengerApiService
import io.reactivex.android.schedulers.AndroidSchedulers
import io.reactivex.schedulers.Schedulers


class ChatInteractorImpl(context: Context) : ChatInteractor {

  private val preferences: AppPreferences = AppPreferences.create(context)
  private val service: MessengerApiService = MessengerApiService
                                       .getInstance()
  private val conversationsRepository: ConversationRepository =
                        ConversationRepositoryImpl(context)
```

The method below will be called to load the messages of a conversation thread:

```
override fun loadMessages(conversationId: Long, listener:
ChatInteractor.onMessageLoadFinishedListener) {
  conversationsRepository.findConversationById(conversationId)
  .subscribeOn(Schedulers.io())
  .observeOn(AndroidSchedulers.mainThread())
  .subscribe({ res -> listener.onLoadSuccess(res)},
  { error ->
    listener.onLoadError()
    error.printStackTrace()})
}
```

The method below will be called to send a message to a user:

```
override fun sendMessage(recipientId: Long, message: String,
listener: ChatInteractor.OnMessageSendFinishedListener) {
service.createMessage(MessageRequestObject(
recipientId, message), preferences.accessToken as String)
.subscribeOn(Schedulers.io())
.observeOn(AndroidSchedulers.mainThread())
.subscribe({ _ -> listener.onSendSuccess()},
{ error ->
  listener.onSendError()
  error.printStackTrace()})
}
}
```

Now, let's handle the `ChatPresenter` and `ChatPresenterImpl` code. For the `ChatPresenter`, we need to create an interface that enforces the declaration of two functions: `sendMessage(Long, String)` and `loadMessages(Long)`. The following is the `ChatPresenter` interface:

```
package com.example.messenger.ui.chat

interface ChatPresenter {

  fun sendMessage(recipientId: Long, message: String)

  fun loadMessages(conversationId: Long)
}
```

The `ChatPresenter` interface's implementation class is shown as follows:

```
package com.iyanuadelekan.messenger.ui.chat

import android.widget.Toast
import com.iyanuadelekan.messenger.data.vo.ConversationVO
import com.iyanuadelekan.messenger.utils.message.Message
import java.text.SimpleDateFormat

class ChatPresenterImpl(val view: ChatView) : ChatPresenter,
        ChatInteractor.OnMessageSendFinishedListener,
        ChatInteractor.onMessageLoadFinishedListener {

    private val interactor: ChatInteractor = ChatInteractorImpl
                                        (view.getContext())

    override fun onLoadSuccess(conversationVO: ConversationVO) {
        val adapter = view.getMessageListAdapter()

        // create date formatter to format createdAt dates
        // received from Messenger API
        val dateFormatter = SimpleDateFormat("yyyy-MM-dd HH:mm:ss")
```

Let's iterate over conversation message loaded from API, and create a new `IMessage` object for message currently iterated upon and add `IMessage` to the start of `MessagesListAdapter`:

```
        conversationVO.messages.forEach { message ->
          adapter.addToStart(Message(message.senderId, message.body,
                dateFormatter.parse(message.createdAt.split(".")[0])), true)
        }
    }

    override fun onLoadError() {
      view.showConversationLoadError()
    }

    override fun onSendSuccess() {
      Toast.makeText(view.getContext(), "Message sent",
    Toast.LENGTH_LONG).show()
    }

    override fun onSendError() {
      view.showMessageSendError()
    }
```

```
    override fun sendMessage(recipientId: Long, message: String) {
      interactor.sendMessage(recipientId, message,this)
    }

    override fun loadMessages(conversationId: Long) {
      interactor.loadMessages(conversationId, this)
    }
  }
```

Keeping with our practice thus far, explanatory comments have been left within the preceding code snippet to aid your understanding.

Last, but not the least, we will work on the ChatActivity. First and foremost, we shall begin by declaring the required properties for our activity and working on its onCreate() lifecycle method.

Modify ChatActivity to contain the following code:

```
package com.example.messenger.ui.chat

import android.content.Context
import android.content.Intent
import android.support.v7.app.AppCompatActivity
import android.os.Bundle
import android.view.MenuItem
import android.widget.Toast
import com.example.messenger.R
import com.example.messenger.data.local.AppPreferences
import com.example.messenger.ui.main.MainActivity
import com.example.messenger.utils.message.Message
import com.stfalcon.chatkit.messages.MessageInput
import com.stfalcon.chatkit.messages.MessagesList
import com.stfalcon.chatkit.messages.MessagesListAdapter
import java.util.*

class ChatActivity : AppCompatActivity(), ChatView,
MessageInput.InputListener {

  private var recipientId: Long = -1
  private lateinit var messageList: MessagesList
  private lateinit var messageInput: MessageInput
  private lateinit var preferences: AppPreferences
  private lateinit var presenter: ChatPresenter
  private lateinit var messageListAdapter: MessagesListAdapter<Message>

  override fun onCreate(savedInstanceState: Bundle?) {
    super.onCreate(savedInstanceState)
```

```
setContentView(R.layout.activity_chat)
supportActionBar?.setDisplayHomeAsUpEnabled(true)
supportActionBar?.title = intent.getStringExtra("RECIPIENT_NAME")

preferences = AppPreferences.create(this)
messageListAdapter = MessagesListAdapter(
preferences.userDetails.id.toString(), null)
presenter = ChatPresenterImpl(this)
bindViews()
```

Let's parse extras bundle from intent which launched the ChatActivity. If either of the extras identified by the keys CONVERSATION_ID and RECIPIENT_ID does not exist, -1 is returned as a default value:

```
val conversationId = intent.getLongExtra("CONVERSATION_ID", -1)
recipientId = intent.getLongExtra("RECIPIENT_ID", -1)
```

If conversationId is not equal to -1, then the conversationId is valid, hence load messages in the conversation:

```
if (conversationId != -1L) {
    presenter.loadMessages(conversationId)
  }
  }
}
```

In the code above, we created recipientId, messageList, messageInput, preferences, presenter, and messageListAdapter properties which are of the type Long, MessageList, MessageInput, AppPreferences, ChatPresenter and MessageListAdapter respectively. messageList is a view which renders distinct views for messages provided to it by messageListAdapter. All the logic contained within onCreate() has to do with the initialization of views within the activity. The code within onCreate() has been commented to give you full understanding of what is going on. Go through each line of the comments patiently before proceeding. ChatActivity implements MessageInput.InputListener. Classes which implement this interface must provide an appropriate onSubmit() method. Let's go ahead and do that. Add the following method to ChatActivity.

Function override from MessageInput.InputListener called when a user submits a message with the MessageInput widget:

```
override fun onSubmit(input: CharSequence?): Boolean {
    // create a new Message object and add it to the
    // start of the MessagesListAdapter
```

```
messageListAdapter.addToStart(Message(
preferences.userDetails.id, input.toString(), Date()), true)

// start message sending procedure with the ChatPresenter
presenter.sendMessage(recipientId, input.toString())

return true
}
```

`onSubmit()` takes a `CharSequence` of the message submitted by the `MessageInput` and creates an appropriate Message instance for it. This instance is then added to the start of the `MessageList` by invoking `messageListAdapter.addToStart()` with the Message instance passed as an argument. After adding the created `Message` to `MessageList`, the `ChatPresenter` instance is used to initialize the sending procedure to the server.

Now let us work on other method overrides we must do. Add the `showConversationLoadError()`, `showMessageSendError()`, `getContext()` and `getMessageListAdapter()` methods shown below to `ChatActivity`:

```
override fun showConversationLoadError() {
  Toast.makeText(this, "Unable to load thread.
  Please try again later.",
    Toast.LENGTH_LONG).show()
}

override fun showMessageSendError() {
  Toast.makeText(this, "Unable to send message.
  Please try again later.",
    Toast.LENGTH_LONG).show()
}

override fun getContext(): Context {
  return this
}

override fun getMessageListAdapter(): MessagesListAdapter<Message> {
  return messageListAdapter
}
```

And finally override `bindViews()`, `onOptionsItemSelected()`, and `onBackPressed()` as follows:

```
override fun bindViews() {
  messageList = findViewById(R.id.messages_list)
  messageInput = findViewById(R.id.message_input)
```

```
    messageList.setAdapter(messageListAdapter)
    messageInput.setInputListener(this)
}

override fun onOptionsItemSelected(item: MenuItem?): Boolean {
    if (item?.itemId == android.R.id.home) {
        onBackPressed()
    }
    return super.onOptionsItemSelected(item)
}

override fun onBackPressed() {
    super.onBackPressed()
    finish()
}
```

So far, so good! You have successfully created the majority of the Messenger app. Go ahead and give yourself a round of applause. The only thing that remains for us to do before wrapping up this chapter is to create a settings activity from which users can update their profile statuses. Feel free to take a well-deserved coffee break before proceeding to the next section.

Creating the application's settings activity

It is now the time to develop a simple application settings activity from which a user can update their profile status. Create a new package within com.example.messenger.ui named settings. Within this package, create a new settings activity. Name the activity SettingsActivity. To create a settings activity, right-click on the settings package, then select **New** | **Activity** | **Settings Activity**. Input the necessary details of the new settings activity, such as an activity name and activity title, then click **Finish**.

In the process of creating a new SettingsActivity, Android Studio will add a number of additional files to your project. In addition to this, a new .xml resource directory (app | res | xml) will be added to your project. This directory should contain the following files:

- pref_data_sync.xml
- pref_general.xml
- pref_headers.xml
- pref_notification.xml

You may choose to delete the `pref_notification.xml` and `pref_data_sync.xml` files. We will not make use of them in this project. Let's take a look at `pref_general.xml`:

```xml
<PreferenceScreen
xmlns:android="http://schemas.android.com/apk/res/android">
  <SwitchPreference
  android:defaultValue="true"
  android:key="example_switch"
  android:summary=
  "@string/pref_description_social_recommendations"
  android:title="@string/pref_title_social_recommendations" />

  <!-- NOTE: EditTextPreference accepts EditText attributes. -->
  <!-- NOTE: EditTextPreference's summary should be set to
  its value by the activity code. -->
  <EditTextPreference
    android:capitalize="words"
    android:defaultValue="@string/pref_default_display_name"
    android:inputType="textCapWords"
    android:key="example_text"
    android:maxLines="1"
    android:selectAllOnFocus="true"
    android:singleLine="true"
    android:title="@string/pref_title_display_name" />

  <!-- NOTE: Hide buttons to simplify the UI.
  Users can touch outside the dialog to
  dismiss it. -->
  <!-- NOTE: ListPreference's summary should be set to
  its value by the activity code. -->
  <ListPreference
    android:defaultValue="-1"
    android:entries="@array/pref_example_list_titles"
    android:entryValues="@array/pref_example_list_values"
    android:key="example_list"
    android:negativeButtonText="@null"
    android:positiveButtonText="@null"
    android:title="@string/pref_title_add_friends_to_messages" />

</PreferenceScreen>
```

The root view of this `xml` layout file is a `PreferenceScreen`. A `PreferenceScreen` is the root of a `Preference` hierarchy. A `PreferenceScreen` in itself is a top-level `Preference`. The word `Preference` has been used a few times now. Let's define what a `Preference` is. It is a representation of the basic `Preference` user-interface building block that is displayed in the form of a list by a `PreferenceActivity`.

The Preference class provides an appropriate view for a preference to be displayed within a PreferenceActivity and its associated SharedPreferences for the storage and retrieval of preference data. SwitchPreference, EditTextPreference, and ListPreference in the preceding code snippet are all subclasses of DialogPreference, which in turn is a subclass of the Preference class. A PreferenceActivity is the base class needed by an activity in order to display a hierarchy of preferences to a user.

The SwitchPreference, EditTextPreference, and ListPreference views in pref_general.xml are not needed. Remove them from the XML file now. We need a preference that enables the user to update their status on our Messenger platform. This is a highly specific use case and thus it comes as no surprise that there's no preference widget that provides us with this ability. No worries! We will implement a custom preference that does the job. Let's call it ProfileStatusPreference. Create a new ProfileStatusPreference class containing the following code in the settings package:

```
package com.example.messenger.ui.settings

import android.content.Context
import android.preference.EditTextPreference
import android.text.TextUtils
import android.util.AttributeSet
import android.widget.Toast
import com.example.messenger.data.local.AppPreferences
import com.example.messenger.data.remote.request.StatusUpdateRequestObject
import com.example.messenger.service.MessengerApiService
import io.reactivex.android.schedulers.AndroidSchedulers
import io.reactivex.schedulers.Schedulers

class ProfileStatusPreference(context: Context, attributeSet: AttributeSet)
  : EditTextPreference(context, attributeSet) {

  private val service: MessengerApiService = MessengerApiService
                                        .getInstance()
  private val preferences: AppPreferences = AppPreferences
                                        .create(context)

  override fun onDialogClosed(positiveResult: Boolean) {
    if (positiveResult) {
```

The following snippet binds `ProfileStatusPreference`'s `EditText` to `etStatus` variable:

```
val etStatus = editText

if (TextUtils.isEmpty(etStatus.text)) {
  // Display error message when user tries
  // to submit an empty status.
  Toast.makeText(context, "Status cannot be empty.",
  Toast.LENGTH_LONG).show()

} else {
  val requestObject =
  StatusUpdateRequestObject(etStatus.text.toString())
```

Let's use `MessengerApiService` to update the user's status:

```
service.updateUserStatus(requestObject,
preferences.accessToken as String)
.subscribeOn(Schedulers.io())
.observeOn(AndroidSchedulers.mainThread())
.subscribe({ res ->
```

Now, we will store the updated user details if status update is successful:

```
preferences.storeUserDetails(res) },
{ error ->
  Toast.makeText(context, "Unable to update status at the " +
  "moment. Try again later.", Toast.LENGTH_LONG).show()
  error.printStackTrace() })
  }
}

super.onDialogClosed(positiveResult)
}
}
```

The `ProfileStatusPreference` class extends `EditTextPreference`.
`EditTextPreference` is a `Preference` that permits string input in
an `EditText`. `EditTextPreference` is a `DialogPreference`, and, as such, presents a
dialog to the user containing the `Preference` view when the `Preference` is clicked. When
a dialog of the `DialogPreference` is closed, its `onDialogClosed(Boolean)` method is
invoked. A positive `Boolean` value argument, `true`, is passed to `onDialogClosed()` when
the dialog is dismissed with a positive result. `false` is passed to `onDialogClosed()` when
the dialog is dismissed with a negative result, for example, when the dialog's cancel button
is clicked.

The `ProfileStatusPreference` overrides the `onDialogClosed()` function of
`EditTextPreference`. If the dialog is closed with a positive result, the validity of the
status contained within the `EditText` function of `ProfileStatusPreference` is checked.
If the status message is valid, the status is updated with the API, otherwise an error
message is shown.

Having created `ProfileStatusPreference`, go back to `pref_general.xml` and update
it to reflect the XML in the following snippet:

```
<PreferenceScreen
xmlns:android="http://schemas.android.com/apk/res/android">
  <com.example.messenger.ui.settings.ProfileStatusPreference
    android:key="profile_status"
    android:singleLine="true"
    android:inputType="text"
    android:maxLines="1"
    android:selectAllOnFocus="true"
    android:title="Profile status"
    android:defaultValue="Available"
    android:summary="Set profile status (visible to contacts)."/>
</PreferenceScreen>
```

As can be seen in the preceding code, we made use of `ProfileStatusPreference` in the
code snippet as we would any other preference bundled within the Android application
framework.

Moving on to other aspects, let's check out `pref_headers.xml`:

```
<preference-headers
xmlns:android="http://schemas.android.com/apk/res/android">
  <!-- These settings headers are only used on tablets. -->
  <header
    android:fragment=
    "com.example.messenger.ui.settings.SettingsActivity
    $GeneralPreferenceFragment"
    android:icon="@drawable/ic_info_black_24dp"
    android:title="@string/pref_header_general" />
  <header
    android:fragment=
    "com.example.messenger.ui.settings.SettingsActivity
    $NotificationPreferenceFragment"
    android:icon="@drawable/ic_notifications_black_24dp"
    android:title="@string/pref_header_notifications" />
  <header
    android:fragment=
    "com.example.messenger.ui.settings.SettingsActivity
    $DataSyncPreferenceFragment"
    android:icon="@drawable/ic_sync_black_24dp"
    android:title="@string/pref_header_data_sync" />
</preference-headers>
```

The preference header file defines headers for various preferences in
the `SettingsActivity`. Modify the file to contain the following code:

```
<preference-headers
xmlns:android="http://schemas.android.com/apk/res/android">
<header
  android:fragment=
  "com.example.messenger.ui.settings.SettingsActivity
  $GeneralPreferenceFragment"
  android:icon="@drawable/ic_info_black_24dp"
  android:title="@string/pref_header_account" />
</preference-headers>
```

Perfectly done! Now we have to work on the `SettingsActivity`. Modify the body
of `SettingsActivity` to contain the content shown in the following code block:

```
package com.example.messenger.ui.settings

import android.content.Intent
import android.os.Bundle
import android.preference.PreferenceActivity
import android.preference.PreferenceFragment
```

```
import android.view.MenuItem
import android.support.v4.app.NavUtils
import com.example.messenger.R
```

`PreferenceActivity` **presenting a set of application settings:**

```
class SettingsActivity : AppCompatPreferenceActivity() {

  override fun onCreate(savedInstanceState: Bundle?) {
    super.onCreate(savedInstanceState)
    supportActionBar?.setDisplayHomeAsUpEnabled(true)
  }

  override fun onMenuItemSelected(featureId: Int, item: MenuItem):
  Boolean {
    val id = item.itemId

    if (id == android.R.id.home) {
      if (!super.onMenuItemSelected(featureId, item)) {
        NavUtils.navigateUpFromSameTask(this)
      }
      return true
    }
    return super.onMenuItemSelected(featureId, item)
  }
```

The following function `onBuildHeaders()` is called when the activity needs a list of headers build:

```
  override fun onBuildHeaders(target: List<PreferenceActivity.Header>)
  {
    loadHeadersFromResource(R.xml.pref_headers, target)
  }
```

The below method preventing fragment injection from malicious applications and all unknown fragments should be denied here:

```
  override fun isValidFragment(fragmentName: String): Boolean {
    return PreferenceFragment::class.java.name == fragmentName
    || GeneralPreferenceFragment::class.java.name == fragmentName
  }
```

The below fragment shows general preferences:

```kotlin
class GeneralPreferenceFragment : PreferenceFragment() {

  override fun onCreate(savedInstanceState: Bundle?) {
    super.onCreate(savedInstanceState)

    addPreferencesFromResource(R.xml.pref_general)
    setHasOptionsMenu(true)
  }

  override fun onOptionsItemSelected(item: MenuItem): Boolean {
    val id = item.itemId

    if (id == android.R.id.home) {
      startActivity(Intent(activity, SettingsActivity::class.java))
      return true
    }
    return super.onOptionsItemSelected(item)
  }
}
```

The `SettingsActivity` extends `AppCompatPreferenceActivity`—an activity that implements the required calls to be used with `AppCompat`. `SettingsActivity` is a `PreferenceActivity` that represents a set of application settings. The `onBuildHeaders()` function of `SettingsActivity` is called when the activity needs a list of headers built. `isValidFragment()` prevents malicious applications from injecting fragments into the `SettingsActivity`. The `isValidFragment()` returns true when a fragment is valid and false otherwise.

Within `SettingsActivity`, we defined a `GeneralPreferenceFragment` class. `GeneralPreferenceFragment` extends `PreferenceFragment`. The `PreferenceFragment` fragment is an abstract class defined in the Android application framework. A `PreferenceFragment` shows a hierarchy of `Preference` instances as lists.

Preferences from `pref_general.xml` are added to the `GeneralPreferenceFragment` in the `onCreate()` method by the invocation of `addPreferencesFromResource(R.xml.pref_general)`.

With these changes made to the `SettingsActivity`, I am pleased to inform you that you have successfully finished work on the settings of the Messenger application.

Having completed the `SettingsActivity`, we are now ready to run the Messenger app. Go ahead and build and run the Messenger application on a device (virtual or physical). Once the app launches, you will be directed straight to the `LoginActivity`.

The first thing we must do is register a new user on the Messenger platform. We can do this on the `SignUpActivity`. Go ahead and click on the **DON'T HAVE AN ACCOUNT? SIGN UP!** button. You will be directed to the `SignUpActivity`:

Create a new user in this activity. Enter popeye as the username, as well as a phone number and a password, then click the **SIGN UP** button. A new user will be registered on the Messenger platform with the username popeye. Once the registration is completed, you will be directed to the MainActivity and the conversations view will be rendered immediately:

As the newly registered user does not have any active conversations, a toast message informing them of this will be displayed. We need to create another user on the messenger platform to demonstrate the chat functionality. Log out of popeye's account by clicking the three dots at the top-right corner of the screen and selecting **logout**:

Once logged out, create a new Messenger account with the username `dexter`. After logging in as `dexter`, click on the new message creation floating action button at the bottom-right of the conversations view. The contacts view will be rendered to you:

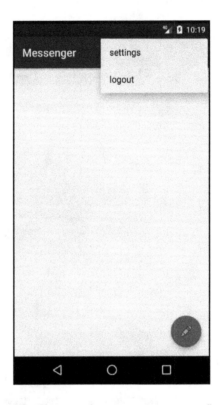

Clicking on the **popeye** contact will open the `ChatActivity`. Let's send a message to `popeye`. Type `Hey Popeye!` into the message input field at the bottom of the screen and click **Send**. The message will immediately be sent to `popeye`:

Upon going back to the conversation view of the `MainActivity`, you will notice a conversation item now exists for the conversation initiated with `popeye`:

Let's check whether the message has actually been delivered to popeye. Log out of the Messenger platform and then log in as popeye. Upon login, you will be greeted by the new conversation initiated by dexter:

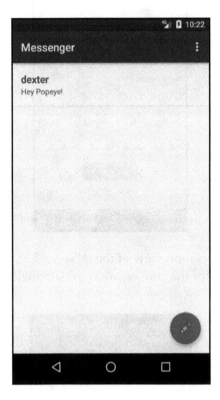

Fantastic! It's been delivered. Now let's reply to `dexter`. Open the conversation and send `dexter` a message:

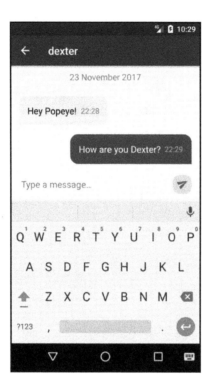

We sent a simple `How are you Dexter?` in the preceding screenshot. It is time to update popeye's profile status. Navigate back to the main activity and access the settings activity (click the three dots on the action bar and select **Settings**). Tapping **Account** in the launched settings activity will display the general preference fragment. Click on the **Profile status** preference:

A dialog containing an `EditText` in which you can input a new profile status will pop up. Input a status message of your choosing and click **OK**:

The status of the current profile will be updated immediately.

At this juncture, I am pleased to inform you that you have successfully implemented the Messenger application in its entirety. Feel free to make modifications and additions to the code we created in this chapter—you will learn a lot more if you do. Before we conclude this chapter, there are two topics we need to cover briefly. The first is application testing, and the second is performing background tasks.

Android application testing

Application testing is the process by which a developed software application is tested to assert its software quality. Many factors contribute to software quality. Such factors include application usability, functionality, reliability, and consistency. A number of advantages arise from testing an Android application. These advantages include but are not limited to:

- Fault detection
- Increased software stability

The integrals of Android application testing span far and wide, and, as such, are beyond the scope of this book. Nevertheless, the following is a list of Android-testing resources you may choose to (and probably should) explore in your free time:

- Espresso
  (https://developer.android.com/training/testing/espresso/index.html)
- Roboelectric (http://robolectric.org)
- Mockito (http://site.mockito.org)
- Calabash (https://github.com/calabash/calabash-android)

Performing background operations

We made use of RxAndroid extensively in the process of developing the Messenger application to perform asynchronous operations. In many cases, when using RxAndroid, we observed the outcome of background operations on the main thread of the Android application. In some cases, you may not want to use a third-party library, such as RxAndroid, to do this. Instead, you may want to use a solution bundled in the Android application framework. Android provides a number of options to achieve this goal. One such option is AsyncTask.

AsyncTask

The AsyncTask class enables the performance of background operations and the publishing of operation results on the application UI thread without the burden of managing handlers and threads. AsyncTask is best used in situations where short operations need to be run. The computations of an AsyncTask run on a background thread and their results are published to the UI thread. You can find out more about AsyncTask here: https://developer.android.com/reference/android/os/AsyncTask.html.

IntentService

An IntentService is a good candidate for performing scheduled operations that run in the background, independent of an activity. As described in the Android developer reference, IntentService is a base class for services that handle asynchronous requests (expressed as Intents) on demand. Clients send requests through startService (Intent) calls; the service is started as needed and handles each Intent in turn using a worker thread, and stops itself when it runs out of work. You can learn more about IntentService here: https://developer.android.com/reference/android/app/IntentService.html.

Summary

In this chapter, we completed the development of the Messenger Android application. In the process of doing so, we learned how to make use of ChatKit—a third-party library for creating beautiful chat user interfaces. In addition to this, we further explored the utilities offered to us by the Android application framework. We got a firsthand look at the development of a settings activity in Android, which helped us to learn about PreferenceScreen, PreferenceActivity, DialogPreference, Preference, and PreferenceFragment. Finally, we briefly discussed Android application testing and performing background operations.

In the next chapter, we will explore the various storage options provided to us by the Android application framework.

7
Storing Information in a Database

In the last chapter, we touched on important topics, such as the use of third party libraries, Android application testing, and how to run background tasks on the Android platform. In this chapter, we will focus on the storage of data. Over the course of this book, we have stored persistent application data at different instances when necessary. So far, we have made use of `SharedPreferences` to cater to all our data storage needs. This is by no means the only option for data storage the Android application framework provides. In this chapter, we will take an in-depth look at the means of data storage available to us on Android. In the process of doing this, we will learn about the following:

- Internal storage
- External storage
- Network storage
- SQLite databases
- Content providers

In addition, we will identify which storage method is best for various use cases. Let's start the chapter off by learning about internal storage.

 Ellipses between code signifies more code that will be present in the code files.

Working with internal storage

This is an available storage medium on the Android application framework that empowers developers to store private data on a device's memory. As the word *private* implies, other applications cannot access the data stored by an app via internal storage. In addition, these files will be removed from storage when that app is uninstalled.

Writing files to internal storage

In order to create a private file on internal storage, it is necessary to call openFileOutput(). The openFileOutput() function takes two arguments. The first is the name (in form of a String) of the file to open and the second is the operating mode. Note that openFileOutput() must be called within an instance of Context, such as an Activity.

The openFileOutput() method returns a FileOutputStream. This can then be used to tell the file it is done with the write() method. Once writing is done, the FileOutputStream should be closed by calling its close() method. The following code snippet demonstrates this process:

```
private fun writeFile(fileName: String) {
    val content: String = "Hello world"
    val stream: FileOutputStream = openFileOutput(fileName,
                                    Context.MODE_PRIVATE)
    stream.write(content.toByteArray())
    stream.close()
}
```

MODE_PRIVATE is an operating mode that creates a file with a given name (or replaces a file with a corresponding name) and makes it private to your application.

Reading files from internal storage

In order to read a private file, get a FileInputStream by calling openFileInput(). This method takes a single argument—the name of the file to be read. openFileInput() must be called within an instance of Context. After retrieving a FileInputStream, you read bytes from the file by calling its read() function. Once you are done reading from the file, close it by invoking close().

Check out the following code:

```
private fun readFile(fileName: String) {
    val stream: FileInputStream = openFileInput(fileName)
    val data = ByteArray(1024)

    stream.read(data)
    stream.close()
}
```

A sample application using internal storage

As a suitable example generally helps the understanding of a concept, let's create a quick bare bones *file-updater* application that utilizes internal storage. The operation of the file updater is simple. It collects text data from the user via an input field and updates a file stored in internal storage. A user can then check the text existing in that file from a view within the application. Simple, right? It's supposed to be! Create a new Android project and name it whatever you want. Just ensure that whatever name you give it reflects the purpose of the application. Once the project is created, create two new packages named base and main, within the src project package.

Add a BaseView interface to the base package, containing the following code:

```
package com.example.storageexamples.base

interface BaseView {

    fun bindViews()

    fun setupInstances()
}
```

You will already be familiar with view interface definitions if you read the previous chapter. If you haven't, this would be a good time to do so. In the main package, create a MainView extending the BaseView as follows:

```
package com.example.storageexamples.main

import com.example.storageexamples.base.BaseView

interface MainView : BaseView {

    fun navigateToHome()
    fun navigateToContent()
```

```
}
```

The file-updater application is going to have two distinct views in the form of fragments that can be shown to a user. The first view is the home view from which a user can update the content of the file, and the second a content view from which the user can read the content of the updated file.

Now, create a new empty activity within `main`. Name this activity `MainActivity`. Ensure that you make `MainActivity` the launcher activity. Once `MainActivity` is created, ensure that it extends the `MainView` before proceeding. Open `activity_main.xml` and edit it to contain the following content:

```
<?xml version="1.0" encoding="utf-8"?>
<android.support.constraint.ConstraintLayout

xmlns:android="http://schemas.android.com/apk/res/android"
 xmlns:tools="http://schemas.android.com/tools"
 android:layout_width="match_parent"
 android:layout_height="match_parent"
 tools:context="com.example.storageexamples.main.MainActivity">

  <LinearLayout
        android:id="@+id/ll_container"
        android:layout_width="match_parent"
        android:layout_height="match_parent"
        android:orientation="vertical"/>
</android.support.constraint.ConstraintLayout>
```

We added a `LinearLayout` view group to the root view of the layout file. This layout is going to serve as the container for the home and content fragment of the application. We now need layouts for the home and content fragments. The following is the home fragment layout (`fragment_home.xml`):

```
<?xml version="1.0" encoding="utf-8"?>
<LinearLayout xmlns:android="http://schemas.android.com/apk/res/android"
 android:orientation="vertical"
 android:layout_width="match_parent"
 android:paddingTop="@dimen/padding_default"
 android:paddingBottom="@dimen/padding_default"
 android:paddingStart="@dimen/padding_default"
 android:paddingEnd="@dimen/padding_default"
 android:gravity="center_horizontal"
 android:layout_height="match_parent">
  <TextView
        android:id="@+id/tv_header"
        android:layout_width="wrap_content"
```

```
            android:layout_height="wrap_content"
            android:text="@string/header_title"
            android:textSize="45sp"
            android:textStyle="bold"/>
    <EditText
            android:id="@+id/et_input"
            android:layout_width="match_parent"
            android:layout_height="wrap_content"
            android:layout_marginTop="@dimen/margin_top_large"
            android:hint="@string/hint_enter_text"/>
    <Button
            android:id="@+id/btn_submit"
            android:layout_width="match_parent"
            android:layout_height="wrap_content"
            android:layout_marginTop="@dimen/margin_default"
            android:text="@string/submit"/>
    <Button
            android:id="@+id/btn_view_file"
            android:layout_width="match_parent"
            android:layout_height="wrap_content"
            android:text="@string/view_file"
            android:background="@android:color/transparent"/>
</LinearLayout>
```

Before you open the design window to view how the layout translates we need to add some value resources. Your project's `strings.xml` file should look similar to the following (with the exception of the `app_name` string resource):

```
<resources>
    <string name="app_name">Storage Examples</string>
    <string name="hint_enter_text">Enter text here...</string>
    <string name="submit">Update file</string>
    <string name="view_file">View file</string>
    <string name="header_title">FILE UPDATER</string>
</resources>
```

In addition, your project should have a `dimens.xml` file containing the following code:

```
<?xml version="1.0" encoding="utf-8"?>
<resources>
    <dimen name="padding_default">16dp</dimen>
    <dimen name="margin_default">16dp</dimen>
    <dimen name="margin_top_large">64dp</dimen>
</resources>
```

Once you are done adding the preceding resources, you can view the layout design window of `fragment_home.xml`:

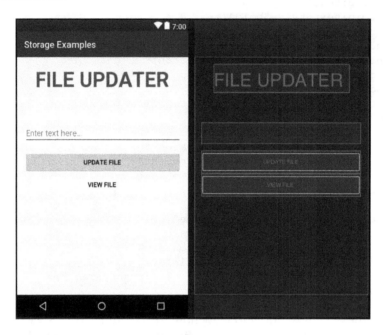

Now for the content fragment layout. Add a `fragment_content.xml` layout file to the layout resource directory containing the following code:

```xml
<?xml version="1.0" encoding="utf-8"?>
<LinearLayout xmlns:android="http://schemas.android.com/apk/res/android"
  android:orientation="vertical"
  android:layout_width="match_parent"
  android:padding="@dimen/padding_default"
  android:layout_height="match_parent">
  <TextView
        android:id="@+id/tv_content"
        android:layout_width="match_parent"
        android:layout_height="wrap_content"
        android:textSize="20sp"
        android:textStyle="bold"
        android:layout_marginTop="@dimen/margin_default"/>
</LinearLayout>
```

The layout contains a single `TextView` that will render the text existing in the internal storage file to the user of the application. You can check the layout design window if you want, but there will not be much to see.

We have reached the point where we need to create appropriate fragment classes to render the fragment layouts we just created. Add the `HomeFragment` class to `MainActivity`, which is as follows:

```
class HomeFragment : Fragment(), BaseView, View.OnClickListener {

    private lateinit var layout: LinearLayout
    private lateinit var tvHeader: TextView
    private lateinit var etInput: EditText
    private lateinit var btnSubmit: Button
    private lateinit var btnViewFile: Button

    private var outputStream: FileOutputStream? = null


    override fun onCreateView(inflater: LayoutInflater,
                              container: ViewGroup?,
                              savedInstanceState: Bundle?): View {

        // inflate the fragment_home.xml layout
        layout = inflater.inflate(R.layout.fragment_home,
                                  container, false) as LinearLayout
        setupInstances()
        bindViews()

        return layout
    }

    override fun bindViews() {
        tvHeader = layout.findViewById(R.id.tv_header)
        etInput = layout.findViewById(R.id.et_input)
        btnSubmit = layout.findViewById(R.id.btn_submit)
        btnViewFile = layout.findViewById(R.id.btn_view_file)

        btnSubmit.setOnClickListener(this)
        btnViewFile.setOnClickListener(this)
    }
```

The following method is for the instantiation of instance properties:

```
    override fun setupInstances() {
```

Let's open a new `FileOutputStream` to a file named `content_file`. This file is a private file stored in internal storage and as such is only accessible by your application:

```
outputStream = activity?.openFileOutput("content_file",
                                    Context.MODE_PRIVATE)
}
```

The following function is called to display an error to the user if an invalid input is given:

```
private fun showInputError() {
    etInput.error = "File input cannot be empty."
    etInput.requestFocus()
}
```

Let's write a string content with the use of file via a `FileOutputStream`

```
private fun writeFile(content: String) {
    outputStream?.write(content.toByteArray())
    outputStream?.close()
}
```

And the function below is called to clear the input in the input field:

```
private fun clearInput() {
    etInput.setText("")
}
```

The following code snippet shows a success message to the user when invoked:

```
private fun showSaveSuccess() {
    Toast.makeText(activity, "File updated successfully.",
                Toast.LENGTH_LONG).show()
}

override fun onClick(view: View?) {
    val id = view?.id

    if (id == R.id.btn_submit) {
        if (TextUtils.isEmpty(etInput.text)) {
```

It will display an error message if the user submits an empty value as file content input:

```
            showInputError()
        } else {
```

Let's write content to the file, clear the input `EditText` and show a file update success message:

```
        writeFile(etInput.text.toString())
        clearInput()
        showSaveSuccess()
    }
} else if (id == R.id.btn_view_file) {
    // retrieve a reference to MainActivity
    val mainActivity = activity as MainActivity
```

Let's navigate user to the content fragment and display home button on action bar to enable a user go back to the previous fragment:

```
        mainActivity.navigateToContent()
        mainActivity.showHomeNavigation()
    }
  }
}
```

Read through the comments of `HomeFragment` to ensure you understand what's going on. Now that we have added `HomeFragment` to `MainActivity`, we must also add a fragment to render the `fragment_content.xml` layout to a user. The following is the necessary `ContentFragment` class. Add this class to `MainActivity`:

```
class ContentFragment : Fragment(), BaseView {

    private lateinit var layout: LinearLayout
    private lateinit var tvContent: TextView

    private lateinit var inputStream: FileInputStream

    override fun onCreateView(inflater: LayoutInflater?,
                              container: ViewGroup?,
                              savedInstanceState: Bundle?): View {

      layout = inflater?.inflate(R.layout.fragment_content,
                                 container, false) as LinearLayout
      setupInstances()
      bindViews()

      return layout
    }

    override fun onResume() {
```

Let's update content rendered in `TextView` upon resumption of fragment:

```
    updateContent()
    super.onResume()
}

private fun updateContent() {
    tvContent.text = readFile()
}

override fun bindViews() {
    tvContent = layout.findViewById(R.id.tv_content)
}

override fun setupInstances() {
    inputStream = activity.openFileInput("content_file")
}
```

The following code reads the content of file in internal storage and returns content as a string:

```
    private fun readFile(): String {
        var c: Int
        var content = ""

        c = inputStream.read()

        while (c != -1) {
            content += Character.toString(c.toChar())
            c = inputStream.read()
        }

        inputStream.close()

        return content
    }
}
```

The `tvContent` instance (the `TextView` displaying the file content to the user) is updated upon resumption of the `ContentFragment` activity. The content of the `TextView` is updated by setting the text of the `TextView` to the content read from the file with `readFile()`. The last thing that is necessary is the completion of `MainActivity`.

Your fully completed `MainActivity` class should look similar to the following:

```
package com.example.storageexamples.main

import android.aupport.v4.app.Fragment
import android.content.Context
import android.support.v7.app.AppCompatActivity
import android.os.Bundle
import android.text.TextUtils
import android.view.LayoutInflater
import android.view.MenuItem
import android.view.View
import android.view.ViewGroup
import android.widget.*
import com.example.storageexamples.R
import com.example.storageexamples.base.BaseView
import java.io.FileInputStream
import java.io.FileOutputStream

class MainActivity : AppCompatActivity(), MainView {

  private lateinit var llContainer: LinearLayout
```

Let's set up fragment instances:

```
    private lateinit var homeFragment: HomeFragment
    private lateinit var contentFragment: ContentFragment

    override fun onCreate(savedInstanceState: Bundle?) {
      super.onCreate(savedInstanceState)
      setContentView(R.layout.activity_main)
      setupInstances()
      bindViews()
      navigateToHome()
    }

    override fun bindViews() {
      llContainer = findViewById(R.id.ll_container)
    }

    override fun setupInstances() {
      homeFragment = HomeFragment()
      contentFragment = ContentFragment()
    }

    private fun hideHomeNavigation() {
```

```kotlin
      supportActionBar?.setDisplayHomeAsUpEnabled(false)
    }

    private fun showHomeNavigation() {
      supportActionBar?.setDisplayHomeAsUpEnabled(true)
    }

    override fun navigateToHome() {
      val transaction = supportFragmentManager.beginTransaction()
      transaction.replace(R.id.ll_container, homeFragment)
      transaction.commit()

      supportActionBar?.title = "Home"
    }

    override fun navigateToContent() {
      val transaction = supportFragmentManager.beginTransaction()
      transaction.replace(R.id.ll_container, contentFragment)
      transaction.commit()

      supportActionBar?.title = "File content"
    }

    override fun onOptionsItemSelected(item: MenuItem?): Boolean {
      val id = item?.itemId

      if (id == android.R.id.home) {
        navigateToHome()
        hideHomeNavigation()
      }

      return super.onOptionsItemSelected(item)
    }

    class HomeFragment : Fragment(), BaseView, View.OnClickListener {

      private lateinit var layout: LinearLayout
      private lateinit var tvHeader: TextView
      private lateinit var etInput: EditText
      private lateinit var btnSubmit: Button
      private lateinit var btnViewFile: Button

      private lateinit var outputStream: FileOutputStream


      override fun onCreateView(inflater: LayoutInflater?,
                                container: ViewGroup?,
                                savedInstanceState: Bundle?): View {
```

Let's create the `fragment_home.xml` layout:

```
    layout = inflater?.inflate(R.layout.fragment_home,
                                container, false) as LinearLayout
    setupInstances()
    bindViews()

    return layout
}

override fun bindViews() {
  tvHeader = layout.findViewById(R.id.tv_header)
  etInput = layout.findViewById(R.id.et_input)
  btnSubmit = layout.findViewById(R.id.btn_submit)
  btnViewFile = layout.findViewById(R.id.btn_view_file)

  btnSubmit.setOnClickListener(this)
  btnViewFile.setOnClickListener(this)
}

//Method for the instantiation of instance properties
override fun setupInstances() {
```

Let's open a new `FileOutputStream` to a file named `content_file`. This file is a private file stored in internal storage and as such is only accessible by your application:

```
    outputStream = activity.openFileOutput("content_file",
                                            Context.MODE_PRIVATE)
}

//Called to display an error to the user if an invalid input is given
private fun showInputError() {
  etInput.error = "File input cannot be empty."
  etInput.requestFocus()
}

// Writes string content to a file via a [FileOutputStream]
private fun writeFile(content: String) {
  outputStream.write(content.toByteArray())
}

//Called to clear the input in the input field

private fun clearInput() {
  etInput.setText("")
}
```

```
//Shows a success message to the user when invoked.
private fun showSaveSuccess() {
  Toast.makeText(activity, "File updated successfully.",
               Toast.LENGTH_LONG).show()
}

override fun onClick(view: View?) {
  val id = view?.id

  if (id == R.id.btn_submit) {
```

The following code snippet displays an error message if the user submits an empty value as file content input:

```
if (TextUtils.isEmpty(etInput.text)) {
  showInputError()
} else {
  //Write content to the file, clear the input
  //EditText and show a file update success message
  writeFile(etInput.text.toString())
  clearInput()
  showSaveSuccess()
}
} else if (id == R.id.btn_view_file) {
// retrieve a reference to MainActivity
val mainActivity = activity as MainActivity
```

We will navigate user to the content fragment and display home button on action bar to enable a user go back to the previous fragment:

```
    mainActivity.navigateToContent()
    mainActivity.showHomeNavigation()
  }
 }
}

class ContentFragment : Fragment(), BaseView {

  private lateinit var layout: LinearLayout
  private lateinit var tvContent: TextView

  private lateinit var inputStream: FileInputStream

  override fun onCreateView(inflater: LayoutInflater?,
                      container: ViewGroup?,
                      savedInstanceState: Bundle?): View {
```

```
    layout = inflater?.inflate(R.layout.fragment_content,
                        container, false) as LinearLayout
    setupInstances()
    bindViews()

    return layout
}
```

Let's update content rendered in `TextView` upon resumption of fragment:

```
override fun onResume() {
  updateContent()
  super.onResume()
}

private fun updateContent() {
  tvContent.text = readFile()
}

override fun bindViews() {
  tvContent = layout.findViewById(R.id.tv_content)
}

override fun setupInstances() {
  inputStream = activity.openFileInput("content_file")
}
```

Now, we will read the content of file in internal storage and return content as a string:

```
    private fun readFile(): String {
      var c: Int
      var content = ""

      c = inputStream.read()

      while (c != -1) {
        content += Character.toString(c.toChar())
        c = inputStream.read()
      }

      inputStream.close()

      return content
    }
  }
}
```

Now that we have completed `MainActivity`, the application is ready to be run.

Build and run the project on a device of your choice. Upon launch of the application, the home fragment of `MainActivity` will be presented on the device's display. Type in any content of your choice within the `EditText` input:

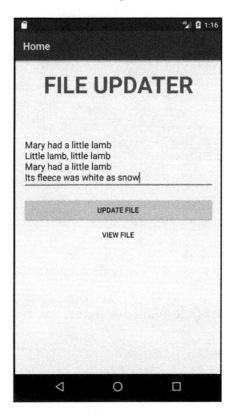

After adding content to the `EditText`, click the **UPDATE FILE** button. The internal storage file will be updated with the content provided, and you will be notified upon completion of this update with a toast message. Having updated the file, click the **VIEW FILE** button:

The content fragment will be displayed with its `TextView` containing the updated file content. Though simple, the file-updater application we just created is a good example showing how Android applications work with internal storage.

Saving cached files

In the event that you would rather not store data permanently, but instead cache data within storage, make use of `cacheDir` to open a file that represents a directory (within internal storage) where your application should save temporary cache files.

`cacheDir` returns a `File`. As such, you may make use of all methods put at your disposal by the `File` class, such as the `outputStream()`, which returns a `FileOutputStream`.

Working with external storage

External storage is used to create and access non-private, shared, world-readable files. Shared external storage is supported by all Android devices. The first thing you need to start using external storage is its permission.

Getting external storage permission

The `READ_EXTERNAL_STORAGE` and `WRITE_EXTERNAL_STORAGE` permissions must be acquired by your application before it can make use of external storage APIs. `READ_EXTERNAL_STORAGE` is necessary when you only need to read from external storage. The `WRITE_EXTERNAL_STORAGE` permission is necessary when your application requires the ability to write directly to external storage.

These two permissions can be added with ease to the manifest file, as we have seen in previous chapters:

```
<uses-permission android:name="android.permission.WRITE_EXTERNAL_STORAGE"
/>
<uses-permission android:name="android.permission.READ_EXTERNAL_STORAGE" />
```

It should be noted that `WRITE_EXTERNAL_STORAGE` implicitly includes `READ_EXTERNAL_STORAGE`. As such, if both permissions are required, it is only necessary to request the `WRITE_EXTERNAL_STORAGE` permission:

```
<uses-permission android:name="android.permission.WRITE_EXTERNAL_STORAGE"
/>
```

Asserting media availability

Sometimes—for one reason or another, such as the storage device being missing—an external storage medium may not be available for access. As a consequence, it is important to check the availability of external storage media before attempting to make use of them.

The `getExternalStorageState()` method should be called in order to check whether media are available. You can use the following code snippet to check whether external storage is available to be written to in your application:

```
private fun isExternalStorageWritable(): Boolean {
  val state = Environment.getExternalStorageState()

  return Environment.MEDIA_MOUNTED == state
}
```

Firstly, we retrieve the current state of external storage and then we check whether it is in the `MEDIA_MOUNTED` state. If it is in this state, your application can write to it. Hence, true is returned by `isExternalStorageWritable()`.

Checking whether external storage is readable is similarly easy:

```
private fun isExternalStorageReadable(): Boolean {
  val state = Environment.getExternalStorageState()

  return Environment.MEDIA_MOUNTED == state ||
       Environment.MEDIA_MOUNTED_READ_ONLY == state
}
```

Your application can read data from external storage if it is in either the `MEDIA_MOUNTED` or the `MEDIA_MOUNTED_READ_ONLY` state.

Storing sharable files

Files that a user or other applications may need to access later on should be stored in a shared public directory. Examples of such directories are the `Pictures/` and `Music/` directories.

The `getExternalStoragePulicDirectory()` method should be called by your application in order to retrieve a `File` representing a required public directory. The type of directory to be retrieved should be passed as the sole argument to the function.

The following is a function that creates a directory for music to be stored:

```
private fun getMusicStorageDir(collectionName: String): File {
  val file = File(Environment.getExternalStoragePublicDirectory(
       Environment.DIRECTORY_MUSIC), collectionName)

  if (!file.mkdir()) {
    Log.d("DIR_CREATION_STATUS", "Directory creation failed.")
```

```
    }

    return file
}
```

In the case where an error occurs when creating the public directory, an appropriate message is logged to the console.

Caching files with external storage

Scenarios may arise in which you need to cache files with external storage. You can open a file representing the external storage directory in which your application should save cached files with `externalCacheDir`.

Network storage

This is the storage of data on a remote server. Unlike other means of storage we have talked about, this kind of storage utilizes a network connection to store and retrieve data that exists on a remote server. You utilized this storage medium when you created the messenger Android application in the previous chapter. The messenger application relied heavily on a remote server for the storage and retrieval of information. The client-server architecture is primarily in play when a remote server acts as a data source for a client application. The client sends a request for the data required (typically a GET request) to the server via HTTP, and the server responds by sending the required data as a response, thus completing the HTTP transaction cycle.

Working with a SQLite database

SQLite is a popular **relational database management system** (**RDBMS**) that, unlike many RDBMS systems, is not a client-server database engine. Instead, a SQLite database is embedded directly into an application.

Android provides full support for SQLite. SQLite databases are accessible throughout an Android project by classes. Note that in Android, a database is only accessible to the application that created it.

The use of the Room persistence library is the recommended method of working with SQLite in Android. The first step to work with `room` in Android is the inclusion of its necessary dependencies in a project's `build.gradle` script:

```
implementation "android.arch.persistence.room:runtime:1.0.0-alpha9-1"
implementation "android.arch.persistence.room:rxjava2:1.0.0-alpha9-1"
implementation "io.reactivex.rxjava2:rxandroid:2.0.1"
kapt "android.arch.persistence.room:compiler:1.0.0-alpha9-1"
```

Entities can easily be created with the help of `Room`. All entities must be annotated with `@Entity`. The following is a simple `User` entity:

```
package com.example.roomexample.data

import android.arch.persistence.room.ColumnInfo
import android.arch.persistence.room.Entity
import android.arch.persistence.room.PrimaryKey

@Entity
data class User(
    @ColumnInfo(name = "first_name")
    var firstName: String = "",
    @ColumnInfo(name = "surname")
    var surname: String = "",
    @ColumnInfo(name = "phone_number")
    var phoneNumber: String = "",
    @PrimaryKey(autoGenerate = true)
    var id: Long = 0
)
```

`Room` will create the necessary SQLite table for the `User` entity defined. The table will have a name, `user`, and will have four attributes: `id`, `first_name`, `surname`, and `phone_number`. The `id` attribute is the primary of the user table created. We specified that this should be the case by using the `@PrimaryKey` annotation. We specified that the primary keys of each record in the user table should be generated by `Room`, by setting `autoGenerate = true` in the `@PrimaryKey` annotation. `@ColumnInfo` is an annotation used to specify additional information pertaining to a column in a table. Take the following code snippet, for example:

```
@ColumnInfo(name = "first_name")
var firstName: String = ""
```

The preceding code specifies that there is a `firstName` attribute possessed by a `User`. `@ColumnInfo(name ="first_name")` sets the name of the column in the user table for the `firstName` attribute to `first_name`.

In order to read and write records to and from a database, you need a **Data Access Object (DAO)**. A DAO allows the performance of database operations with the use of annotated methods. The following is a DAO for the `User` entity:

```
package com.example.roomexample.data

import android.arch.persistence.room.Dao
import android.arch.persistence.room.Insert
import android.arch.persistence.room.OnConflictStrategy
import android.arch.persistence.room.Query
import io.reactivex.Flowable

@Dao
interface UserDao {

    @Query("SELECT * FROM user")
    fun all(): Flowable<List<User>>

    @Query("SELECT * FROM user WHERE id = :id")
    fun findById(id: Long): Flowable<User>

    @Insert(onConflict = OnConflictStrategy.REPLACE)
    fun insert(user: User)
}
```

The `@Query` annotation marks a method in a DAO class as a query method. The query that will be run when the method is invoked is passed as a value to the annotation. Naturally, the queries passed to `@Query` are SQL queries. The writing of SQL queries is far too vast a topic to cover here, but it is a good idea to take some time to understand how to write them properly.

The `@Insert` annotation is used to insert data into a table. Other important annotations that exist are `@Update` and `@Delete`. They are used to update and delete data within a database table.

Lastly, after creating the necessary entities and DAOs, you must define your application's database. In order to do this, you must create a subclass of `RoomDatabase` and annotate it with `@Database`. At the minimum, the annotation should include a collection of entity class references and a database version number. The following is a sample `AppDatabase` abstract class:

```
@Database(entities = [User::class], version = 1)
public abstract class AppDatabase : RoomDatabase() {
    abstract fun userDao(): UserDao
}Now that we have our DAO and entity created, we must create an AppDatabase
```

```
class. Add
```

Once your app database class is created, you can get an instance of the database by calling `databaseBuilder()`:

```
val db = Room.databaseBuilder(<context>, AppDatabase::class.java,
                             "app-database").build()
```

Once you have an instance of your `RoomDatabase`, you can use it to retrieve data access objects that in turn can be used to read, write, update, query, and delete data from a database.

Keeping with the practice we have adopted so far, let's create a simple application that shows how to utilize SQLite with the help of `Room` in Android. The application we are about to build will let an app user manually input information pertaining to people (users), and view all user information input at a later time.

Create a new Android project with an empty `MainActivity` set as its launcher activity. Add the following dependencies to your application's `build.gradle` script:

```
implementation 'com.android.support:design:26.1.0'
implementation "android.arch.persistence.room:runtime:1.0.0"
implementation "android.arch.persistence.room:rxjava2:1.0.0"
implementation "io.reactivex.rxjava2:rxandroid:2.0.1"
kapt "android.arch.persistence.room:compiler:1.0.0"
```

In addition, apply the `kotlin-kapt` standalone plugin to your `build.gradle` script:

```
apply plugin: 'kotlin-kapt'
```

Having added the preceding project dependencies, create a `data` and a `ui` package within the project source package. In the `ui` package, add the `MainView`, which is as follows:

```
package com.example.roomexample.ui

interface MainView {

  fun bindViews()
  fun setupInstances()
}
```

After adding `MainView` to the `ui` package, relocate `MainActivity` to the `ui` package as well. Now let's work on the database of our application. As the application is going to be storing user information, we need to create a `User` entity. Add the following `User` entity to the `data` package:

```
package com.example.roomexample.data

import android.arch.persistence.room.ColumnInfo
import android.arch.persistence.room.Entity
import android.arch.persistence.room.PrimaryKey

@Entity
data class User(
  @ColumnInfo(name = "first_name")
  var firstName: String = "",
  @ColumnInfo(name = "surname")
  var surname: String = "",
  @ColumnInfo(name = "phone_number")
  var phoneNumber: String = "",
  @PrimaryKey(autoGenerate = true)
  var id: Long = 0
)
```

Now create a `UserDao` within the `data` package:

```
package com.example.roomexample.data

import android.arch.persistence.room.Dao
import android.arch.persistence.room.Insert
import android.arch.persistence.room.OnConflictStrategy
import android.arch.persistence.room.Query
import io.reactivex.Flowable

@Dao
interface UserDao {

  @Query("SELECT * FROM user")
  fun all(): Flowable<List<User>>

  @Query("SELECT * FROM user WHERE id = :id")
  fun findById(id: Long): Flowable<User>

  @Insert(onConflict = OnConflictStrategy.REPLACE)
  fun insert(user: User)
}
```

The `UserDao` interface has three methods: `all()`, `findById()`, and `Insert()`. The `all()` method returns a `Flowable` containing a list of all users. The `findById()` method finds a `User` who has an id matching that which is passed to the method, if any, and returns the `User` in a `Flowable`. The `insert()` method is used to insert a user as a record into the `user` table.

Now that we have our DAO and entity created, we must create an `AppDatabase` class. Add the following to the `data` package:

```
package com.example.roomexample.data

import android.arch.persistence.room.Database
import android.arch.persistence.room.Room
import android.arch.persistence.room.RoomDatabase
import android.content.Context

@Database(entities = arrayOf(User::class), version = 1, exportSchema =
false)
internal abstract class AppDatabase : RoomDatabase() {

  abstract fun userDao(): UserDao

  companion object Factory {
    private var appDatabase: AppDatabase? = null

    fun create(ctx: Context): AppDatabase {
      if (appDatabase == null) {
        appDatabase = Room.databaseBuilder(ctx.applicationContext,
                                  AppDatabase::class.java,
                                  "app-database").build()

      }

      return appDatabase as AppDatabase
    }
  }
}
```

We created a `Factory` companion object possessing a single `create()` function that has the sole job of creating an `AppDatabase` instance—if not previously created—and returning that instance for use.

Creating an `AppDatabase` is the last thing we need to do pertaining to data. Now we must create suitable layouts for our application views. We will make use of two fragments in our `MainActivity`. The first will be used to collect input for a new user to be created and the second will display the information of all created users in a `RecyclerView`. Firstly, modify the `activity_main.xml` layout to contain the following:

```xml
<?xml version="1.0" encoding="utf-8"?>
<android.support.constraint.ConstraintLayout
 xmlns:android="http://schemas.android.com/apk/res/android"
 xmlns:tools="http://schemas.android.com/tools"
 android:layout_width="match_parent"
 android:layout_height="match_parent"
 tools:context="com.example.roomexample.ui.MainActivity">

    <LinearLayout
        android:id="@+id/ll_container"
        android:layout_width="match_parent"
        android:layout_height="match_parent"
        android:orientation="vertical"/>
</android.support.constraint.ConstraintLayout>
```

The `LinearLayout` in `activity_main.xml` will contain the fragments of `MainActivity`. Add a `fragment_create_user.xml` file to the `resource` layout directory with the following content:

```xml
<?xml version="1.0" encoding="utf-8"?>
<LinearLayout xmlns:android="http://schemas.android.com/apk/res/android"
 android:orientation="vertical"
 android:layout_width="match_parent"
 android:layout_height="match_parent"
 android:gravity="center_horizontal"
 android:padding="@dimen/padding_default">
  <TextView
        android:layout_width="wrap_content"
        android:layout_height="wrap_content"
        android:textSize="32sp"
        android:text="@string/create_user"/>
    <EditText
        android:id="@+id/et_first_name"
        android:layout_width="match_parent"
        android:layout_height="wrap_content"
        android:layout_marginTop="@dimen/margin_default"
        android:hint="@string/first_name"
        android:inputType="text"/>
    <EditText
        android:id="@+id/et_surname"
```

```
            android:layout_width="match_parent"
            android:layout_height="wrap_content"
            android:layout_marginTop="@dimen/margin_default"
            android:hint="@string/surname"
            android:inputType="text"/>
    <EditText
            android:id="@+id/et_phone_number"
            android:layout_width="match_parent"
            android:layout_height="wrap_content"
            android:layout_marginTop="@dimen/margin_default"
            android:hint="@string/phone_number"
            android:inputType="phone"/>
    <Button
            android:id="@+id/btn_submit"
            android:layout_width="match_parent"
            android:layout_height="wrap_content"
            android:layout_marginTop="@dimen/margin_default"
            android:text="@string/submit"/>
    <Button
            android:id="@+id/btn_view_users"
            android:layout_width="match_parent"
            android:layout_height="wrap_content"
            android:layout_marginTop="@dimen/margin_default"
            android:text="@string/view_users"/>
</LinearLayout>
```

Now add a `fragment_list_users.xml` layout resource with the following content:

```
<?xml version="1.0" encoding="utf-8"?>
<LinearLayout xmlns:android="http://schemas.android.com/apk/res/android"
 android:orientation="vertical"
 android:layout_width="match_parent"
 android:layout_height="match_parent">
  <android.support.v7.widget.RecyclerView
            android:id="@+id/rv_users"
            android:layout_width="match_parent"
            android:layout_height="match_parent"/>
</LinearLayout>
```

The `fragment_list_users.xml` file has a `RecyclerView` that will display the information of each user saved to the database. We must create a view holder layout resource item for this `RecyclerView`. We'll call this layout file `vh_user.xml`. Create a new `vh_user.xml` resource file and add the following content to it:

```
<?xml version="1.0" encoding="utf-8"?>
<LinearLayout xmlns:android="http://schemas.android.com/apk/res/android"
 android:orientation="vertical"
```

```
  android:layout_width="match_parent"
  android:padding="@dimen/padding_default"
  android:layout_height="wrap_content">
   <TextView
        android:id="@+id/tv_first_name"
        android:layout_width="wrap_content"
        android:layout_height="wrap_content"/>
   <TextView
        android:id="@+id/tv_surname"
        android:layout_width="wrap_content"
        android:layout_height="wrap_content"
        android:layout_marginTop="@dimen/margin_default"/>
   <TextView
        android:id="@+id/tv_phone_number"
        android:layout_width="wrap_content"
        android:layout_height="wrap_content"
        android:layout_marginTop="@dimen/margin_default"/>
   <View
        android:layout_width="match_parent"
        android:layout_height="1dp"
        android:layout_marginTop="@dimen/margin_default"
        android:background="#e8e8e8"/>
</LinearLayout>
```

As you might have expected, we must add some string and dimension resources to our project. Open your application's `strings.xml` layout file and add the following string resources to it:

```
<resources>
...
  <string name="first_name">First name</string>
  <string name="surname">Surname</string>
  <string name="phone_number">Phone number</string>
  <string name="submit">Submit</string>
  <string name="create_user">Create User</string>
  <string name="view_users">View users</string>
</resources>
```

Now create the following dimension resources in your project:

```
<?xml version="1.0" encoding="utf-8"?>
<resources>
  <dimen name="padding_default">16dp</dimen>
  <dimen name="margin_default">16dp</dimen>
</resources>
```

It is now time to work on `MainActivity`. As we previously established, we will be using two distinct fragments in our `MainActivity` class. The first fragment enables a person to save the data of an individual to a SQL database, and the next will allow a user to view the information of people that has been saved in the database.

We will create a `CreateUserFragment` first. Add the following fragment class to `MainActivity` (located in the `MainActivity.kt`).

```kotlin
class CreateUserFragment : Fragment(), MainView, View.OnClickListener {

    private lateinit var btnSubmit: Button
    private lateinit var etSurname: EditText
    private lateinit var btnViewUsers: Button
    private lateinit var layout: LinearLayout
    private lateinit var etFirstName: EditText
    private lateinit var etPhoneNumber: EditText

    private lateinit var userDao: UserDao
    private lateinit var appDatabase: AppDatabase

    override fun onCreateView(inflater: LayoutInflater,
                container: ViewGroup?, savedInstanceState: Bundle?): View {
        layout = inflater.inflate(R.layout.fragment_create_user,
                        container, false) as LinearLayout
        bindViews()
        setupInstances()
        return layout
    }

    override fun bindViews() {
        btnSubmit = layout.findViewById(R.id.btn_submit)
        btnViewUsers = layout.findViewById(R.id.btn_view_users)
        etSurname = layout.findViewById(R.id.et_surname)
        etFirstName = layout.findViewById(R.id.et_first_name)
        etPhoneNumber = layout.findViewById(R.id.et_phone_number)

        btnSubmit.setOnClickListener(this)
        btnViewUsers.setOnClickListener(this)
    }

    override fun setupInstances() {
        appDatabase = AppDatabase.create(activity)
            // getting an instance of AppDatabase
        userDao = appDatabase.userDao() // getting an instance of UserDao
    }
```

The following method validates the inputs submitted in the create a user form:

```kotlin
private fun inputsValid(): Boolean {
  var inputValid = true
  val firstName = etFirstName.text
  val surname = etSurname.text
  val phoneNumber = etPhoneNumber.text

  if (TextUtils.isEmpty(firstName)) {
    etFirstName.error = "First name cannot be empty"
    etFirstName.requestFocus()
    inputValid = false

  } else if (TextUtils.isEmpty(surname)) {
    etSurname.error = "Surname cannot be empty"
    etSurname.requestFocus()
    inputValid = false

  } else if (TextUtils.isEmpty(phoneNumber)) {
    etPhoneNumber.error = "Phone number cannot be empty"
    etPhoneNumber.requestFocus()
    inputValid = false

  } else if (!android.util.Patterns.PHONE
                    .matcher(phoneNumber).matches()) {
    etPhoneNumber.error = "Valid phone number required"
    etPhoneNumber.requestFocus()
    inputValid = false
  }

  return inputValid
}
```

The following function shows toast message indicating the user successfully created:

```kotlin
private fun showCreationSuccess() {
  Toast.makeText(activity, "User successfully created.",
                  Toast.LENGTH_LONG).show()
}

override fun onClick(view: View?) {
  val id = view?.id

  if (id == R.id.btn_submit) {
    if (inputsValid()) {
      val user = User(
        etFirstName.text.toString(),
        etSurname.text.toString(),
```

```
                etPhoneNumber.text.toString())

           Observable.just(userDao)
                      .subscribeOn(Schedulers.io())
                      .subscribe( { dao ->
                dao.insert(user)  // using UserDao to save user to database.
                activity?.runOnUiThread { showCreationSuccess() }
              }, Throwable::printStackTrace)
         }
       } else if (id == R.id.btn_view_users) {
         val mainActivity = activity as MainActivity

         mainActivity.navigateToList()
         mainActivity.showHomeButton()
       }
     }
   }
```

We have worked with fragments numerous times already. As such, our focus of
explanation will be on the parts of this fragment that work with the `AppDatabase`. In
`setupInstances()`, we set up references to the `AppDatabase` and `UserDao`. We retrieved
an instance of the `AppDatabase` by invoking the `create()` function of the `Factory`
companion object of the `AppDatabase`. An instance of `UserDao` was easily retrieved by
calling the `appDatabase.userDao()`.

Let's move on to the `onClick()` method of the fragment class. When the submit button is
clicked, the submitted user information will be checked for validity. An appropriate error
message is shown if any of the input is invalid. Once it is asserted that all input are valid, a
new `User` object containing the submitted user information is created and saved to the
database. This is done in the following lines:

```
if (inputsValid()) {
  val user = User(
    etFirstName.text.toString(),
    etSurname.text.toString(),
    etPhoneNumber.text.toString())

  Observable.just(userDao)
            .subscribeOn(Schedulers.io())
            .subscribe( { dao ->
    dao.insert(user)  // using UserDao to save user to database.
    activity?.runOnUiThread { showCreationSuccess() }
  }, Throwable::printStackTrace)
}
```

Creating the `ListUsersFragment` is similarly easy to achieve. Add the following `ListUsersFragment` to `MainActivity`:

```
class ListUsersFragment : Fragment(), MainView {

    private lateinit var layout: LinearLayout
    private lateinit var rvUsers: RecyclerView

    private lateinit var appDatabase: AppDatabase

    override fun onCreateView(inflater: LayoutInflater,
                container: ViewGroup?, savedInstanceState: Bundle?): View {

        layout = inflater.inflate(R.layout.fragment_list_users,
                                    container, false) as LinearLayout
        bindViews()
        setupInstances()

        return layout
    }
```

Bind the user recycler view instance to its layout element:

```
    override fun bindViews() {
        rvUsers = layout.findViewById(R.id.rv_users)
    }

    override fun setupInstances() {
        appDatabase = AppDatabase.create(activity)
        rvUsers.layoutManager = LinearLayoutManager(activity)
        rvUsers.adapter = UsersAdapter(appDatabase)
    }

    private class UsersAdapter(appDatabase: AppDatabase) :
                RecyclerView.Adapter<UsersAdapter.ViewHolder>() {

        private val users: ArrayList<User> = ArrayList()
        private val userDao: UserDao = appDatabase.userDao()

        init {
            populateUsers()
        }

        override fun onCreateViewHolder(parent: ViewGroup?, viewType: Int):
                    ViewHolder {
            val layout = LayoutInflater.from(parent?.context)
                                    .inflate(R.layout.vh_user, parent, false)
```

```
        return ViewHolder(layout)
    }

    override fun onBindViewHolder(holder: ViewHolder?, position: Int) {
      val layout = holder?.itemView
      val user = users[position]

      val tvFirstName = layout?.findViewById<TextView>(R.id.tv_first_name)
      val tvSurname = layout?.findViewById<TextView>(R.id.tv_surname)
      val tvPhoneNumber = layout?.findViewById<TextView>
                          (R.id.tv_phone_number)

      tvFirstName?.text = "First name: ${user.firstName}"
      tvSurname?.text = "Surname: ${user.surname}"
      tvPhoneNumber?.text = "Phone number: ${user.phoneNumber}"
    }

  //Populates users ArrayList with User objects
  private fun populateUsers() {
    users.clear()
```

Let's get all users in the user table of the database. And upon successful retrieval of the list, add all user objects in the list to users `ArrayList`:

```
        userDao.all()
              .subscribeOn(Schedulers.io())
              .observeOn(AndroidSchedulers.mainThread())
              .subscribe({ res ->
          users.addAll(res)
          notifyDataSetChanged()
        }, Throwable::printStackTrace)
    }

    override fun getItemCount(): Int {
      return users.size
    }

    class ViewHolder(itemView: View) : RecyclerView.ViewHolder(itemView)
  }
}
```

UsersAdapter in the ListUsersFragment uses an instance of UserDao to populate its user list. This population is done within populateUsers(). When populateUsers() is invoked, a list of all users that have been saved by the application is retrieved by invoking userDao.all(). Upon successful retrieval of all users, all User objects are added to the users ArrayList of UserAdapter. The adapter is then notified of the change in data in its dataset by the call to notifyDataSetChanged().

MainActivity itself requires some minor additions. Your completed MainActivity should look like this:

```
package com.example.roomexample.ui

import android.app.Fragment
import android.support.v7.app.AppCompatActivity
import android.os.Bundle
import android.support.v7.widget.LinearLayoutManager
import android.support.v7.widget.RecyclerView
import android.text.TextUtils
import android.view.LayoutInflater
import android.view.MenuItem
import android.view.View
import android.view.ViewGroup
import android.widget.*
import com.example.roomexample.R
import com.example.roomexample.data.AppDatabase
import com.example.roomexample.data.User
import com.example.roomexample.data.UserDao
import io.reactivex.Observable
import io.reactivex.android.schedulers.AndroidSchedulers
import io.reactivex.schedulers.Schedulers

class MainActivity : AppCompatActivity() {

  override fun onCreate(savedInstanceState: Bundle?) {
    super.onCreate(savedInstanceState)
    setContentView(R.layout.activity_main)
    navigateToForm()
  }

  private fun showHomeButton() {
    supportActionBar?.setDisplayHomeAsUpEnabled(true)
  }

  private fun hideHomeButton() {
    supportActionBar?.setDisplayHomeAsUpEnabled(false)
  }
```

```
private fun navigateToForm() {
  val transaction = fragmentManager.beginTransaction()
  transaction.add(R.id.ll_container, CreateUserFragment())
  transaction.commit()
}
```

The following function is called when the user click the back button, if the fragments back stack has one or more fragments, the fragments manager pops the fragment and displays it to the user:

```
override fun onBackPressed() {
  if (fragmentManager.backStackEntryCount > 0) {
    fragmentManager.popBackStack()
    hideHomeButton()
  } else {
    super.onBackPressed()
  }
}

private fun navigateToList() {
  val transaction = fragmentManager.beginTransaction()
  transaction.replace(R.id.ll_container, ListUsersFragment())
  transaction.addToBackStack(null)
  transaction.commit()
}

override fun onOptionsItemSelected(item: MenuItem?): Boolean {
  val id = item?.itemId

  if (id == android.R.id.home) {
    onBackPressed()
    hideHomeButton()
  }

  return super.onOptionsItemSelected(item)
}

class CreateUserFragment : Fragment(), MainView, View.OnClickListener {
  ...
}

class ListUsersFragment : Fragment(), MainView {
  ...
}
}
```

We must now run the application to see if it works as we would like. Build and run the project on a device of your choice. Once the project launches, you will come face to face with the user creation form. Go ahead and input some user information in the form:

Once you have input valid information into the create user form, click the **Submit** button to save the user to the application's SQLite database. You will be notified once the user has been saved successfully. Having been notified, click **VIEW USERS** to see the information of the user you just saved:

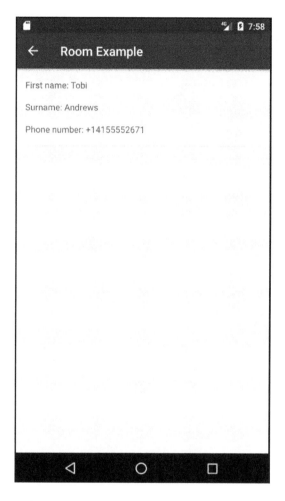

You can create and view information for as many users as you like. There's no upper limit to the amount of information the database can contain!

Working with content providers

We spoke briefly about content providers as Android components in Chapter 2, *Building an Android Application - Tetris*. While doing so, we established the fact that content providers help an application control access to data resources stored either within the application or within another app. In addition, we established that a content provider facilitates the sharing of data with another application via an exposed application programming interface.

A content provider behaves in a way that is similar to the behavior of a database. A content provider permits the insertion, deletion, editing, updating, and querying of content. These abilities are permitted by the use of methods such as insert(), update(), delete(), and query(). In many cases, data controlled by a content provider exists in a SQLite database.

A content provider for your application can be created in five easy steps:

1. Create a content provider class that extends ContentProvider.
2. Define a content URI address.
3. Create a datasource that the content provider will interact with. This datasource is usually in the form of a SQLite database. In cases where SQLite is the datasource, you will need to create a SQLiteOpenHelper and override its onCreate() in order to create the database that will be controlled by the content provider.
4. Implement the required content provider methods.
5. Register the content provider in your project's manifest file.

In all, there are six methods that must be implemented by a content provider. These are:

- onCreate(): This method is called to initialize the database
- query(): This method returns data to the caller via a Cursor
- insert(): This method is called to insert new data into the content provider
- delete(): This method is called to delete data from the content provider
- update(): This method is called to update data in the content provider
- getType(): This method returns the MIME type of data in the content provider when called

In order to ensure you fully understand the workings of a content provider, let's create a quick example project that utilizes a content provider and a SQLite database. Create a new Android studio project named `ContentProvider` and add an empty `MainActivity` to it upon creation. Similar to all other applications created in this chapter, this example is simple in nature. The application allows a user to enter the details of a product (a product name and its manufacturer) in text fields and save them to a SQLite database. The user can then view the information of products that they previously saved with the click of a button.

Modify `activity_main.xml` to contain the following XML:

```xml
<?xml version="1.0" encoding="utf-8"?>
<LinearLayout xmlns:android="http://schemas.android.com/apk/res/android"
 xmlns:tools="http://schemas.android.com/tools"
 android:layout_width="match_parent"
 android:layout_height="match_parent"
 android:orientation="vertical"
 android:gravity="center_horizontal"
 android:padding="16dp"
 tools:context="com.example.contentproviderexample.MainActivity">
 <TextView
        android:layout_width="wrap_content"
        android:layout_height="wrap_content"
        android:gravity="center"
        android:text="@string/content_provider_example"
        android:textColor="@color/colorAccent"
        android:textSize="32sp"/>
 <EditText
        android:id="@+id/et_product_name"
        android:layout_width="match_parent"
        android:layout_height="wrap_content"
        android:layout_marginTop="16dp"
        android:hint="Product Name"/>
 <EditText
        android:id="@+id/et_product_manufacturer"
        android:layout_width="match_parent"
        android:layout_height="wrap_content"
        android:layout_marginTop="16dp"
        android:hint="Product Manufacturer"/>
 <Button
        android:id="@+id/btn_add_product"
        android:layout_width="match_parent"
        android:layout_height="wrap_content"
        android:layout_marginTop="16dp"
        android:text="Add product"/>
 <Button
        android:id="@+id/btn_show_products"
```

```
            android:layout_width="match_parent"
            android:layout_height="wrap_content"
            android:layout_marginTop="16dp"
            android:text="Show products"/>
</LinearLayout>
```

After making the preceding modifications, add the following string resource to your project's `strings.xml` file:

```
<string name="content_provider_example">Content Provider Example</string>
```

Now create a `ProductProvider.kt` file in the `com.example.contentproviderexample` package and add the following content:

```
package com.example.contentproviderexample

import android.content.*
import android.database.Cursor
import android.database.SQLException
import android.database.sqlite.SQLiteDatabase
import android.database.sqlite.SQLiteOpenHelper
import android.database.sqlite.SQLiteQueryBuilder
import android.net.Uri
import android.text.TextUtils

internal class ProductProvider : ContentProvider() {

  companion object {

    val PROVIDER_NAME: String = "com.example.contentproviderexample
                                .ProductProvider"
    val URL: String = "content://$PROVIDER_NAME/products"
    val CONTENT_URI: Uri = Uri.parse(URL)

    val PRODUCTS = 1
    val PRODUCT_ID = 2

    // Database and table property declarations
    val DATABASE_VERSION = 1
    val DATABASE_NAME = "Depot"
    val PRODUCTS_TABLE_NAME = "products"

    // 'products' table column name declarations
    val ID: String = "id"
    val NAME: String = "name"
    val MANUFACTURER: String = "manufacturer"
    val uriMatcher: UriMatcher = UriMatcher(UriMatcher.NO_MATCH)
```

```
val PRODUCTS_PROJECTION_MAP: HashMap<String, String> = HashMap()
```

SQLiteOpenHelper class that creates the content provider's database:

```
private class DatabaseHelper(context: Context) :
    SQLiteOpenHelper(context, DATABASE_NAME, null,
DATABASE_VERSION) {

    override fun onCreate(db: SQLiteDatabase) {
      val query = " CREATE TABLE " + PRODUCTS_TABLE_NAME +
                " (id INTEGER PRIMARY KEY AUTOINCREMENT, " +
                " name VARCHAR(255) NOT NULL, " +
                " manufacturer VARCHAR(255) NOT NULL);"

      db.execSQL(query)
    }

    override fun onUpgrade(db: SQLiteDatabase, oldVersion: Int,
                           newVersion: Int) {
      val query = "DROP TABLE IF EXISTS $PRODUCTS_TABLE_NAME"

      db.execSQL(query)
      onCreate(db)
    }
  }
}

private lateinit var db: SQLiteDatabase

override fun onCreate(): Boolean {
  uriMatcher.addURI(PROVIDER_NAME, "products", PRODUCTS)
  uriMatcher.addURI(PROVIDER_NAME, "products/#", PRODUCT_ID)

  val helper = DatabaseHelper(context)
```

Let's use the SQLiteOpenHelper to get a writable database; a new database is created if one does not already exist:

```
db = helper.writableDatabase

  return true
}

override fun insert(uri: Uri, values: ContentValues): Uri {
  //Insert a new product record into the products table
```

```
    val rowId = db.insert(PRODUCTS_TABLE_NAME, "", values)

    //If rowId is greater than 0 then the product record was added
successfully.
    if (rowId > 0) {
        val _uri = ContentUris.withAppendedId(CONTENT_URI, rowId)
        context.contentResolver.notifyChange(_uri, null)

        return _uri
    }

    // throws an exception if the product was not successfully added.
    throw SQLException("Failed to add product into " + uri)
}

override fun query(uri: Uri, projection: Array<String>?,
                   selection: String?, selectionArgs: Array<String>?,
                   sortOrder: String): Cursor {

    val queryBuilder = SQLiteQueryBuilder()
    queryBuilder.tables = PRODUCTS_TABLE_NAME

    when (uriMatcher.match(uri)) {
        PRODUCTS -> queryBuilder.setProjectionMap(PRODUCTS_PROJECTION_MAP)
        PRODUCT_ID -> queryBuilder.appendWhere(
            "$ID = ${uri.pathSegments[1]}"
        )
    }

    val cursor: Cursor = queryBuilder.query(db, projection, selection,
                         selectionArgs, null, null, sortOrder)

    cursor.setNotificationUri(context.contentResolver, uri)
    return cursor
}

override fun delete(uri: Uri, selection: String,
                    selectionArgs: Array<String>): Int {

    val count = when(uriMatcher.match(uri)) {

        PRODUCTS -> db.delete(PRODUCTS_TABLE_NAME, selection, selectionArgs)
        PRODUCT_ID -> {
            val id = uri.pathSegments[1]
            db.delete(PRODUCTS_TABLE_NAME, "$ID = $id " +
                if (!TextUtils.isEmpty(selection)) "AND
                    ($selection)" else "", selectionArgs)
        }
```

```
      else -> throw IllegalArgumentException("Unknown URI: $uri")
    }

    context.contentResolver.notifyChange(uri, null)
    return count
  }

  override fun update(uri: Uri, values: ContentValues, selection: String,
                      selectionArgs: Array<String>): Int {

    val count = when(uriMatcher.match(uri)) {
      PRODUCTS -> db.update(PRODUCTS_TABLE_NAME, values,
                            selection, selectionArgs)
      PRODUCT_ID -> {
        db.update(PRODUCTS_TABLE_NAME, values,
                  "$ID = ${uri.pathSegments[1]} " +
                  if (!TextUtils.isEmpty(selection)) " AND
                  ($selection)" else "", selectionArgs)
      }
      else -> throw  IllegalArgumentException("Unknown URI: $uri")
    }

    context.contentResolver.notifyChange(uri, null)
    return count
  }

  override fun getType(uri: Uri): String {
    //Returns the appropriate MIME type of records

    return when (uriMatcher.match(uri)){
      PRODUCTS -> "vnd.android.cursor.dir/vnd.example.products"
      PRODUCT_ID -> "vnd.android.cursor.item/vnd.example.products"
      else -> throw IllegalArgumentException("Unpermitted URI: " + uri)
    }
  }
}
```

Having added a suitable `ProductProvider` to provide content pertaining to saved products, we must register the new component in the `AndroidManifest` file. We have added the provider to the manifest file in the following code snippet:

```
<?xml version="1.0" encoding="utf-8"?>
<manifest xmlns:android="http://schemas.android.com/apk/res/android"
    package="com.example.contentproviderexample">

  <application
        android:allowBackup="true"
```

```
        android:icon="@mipmap/ic_launcher"
        android:label="@string/app_name"
        android:roundIcon="@mipmap/ic_launcher_round"
        android:supportsRtl="true"
        android:theme="@style/AppTheme">
    <activity android:name=".MainActivity">
      <intent-filter>
        <action android:name="android.intent.action.MAIN" />

        <category android:name="android.intent.category.LAUNCHER" />
      </intent-filter>
    </activity>
    <provider android:authorities="com.example.contentproviderexample
                    .ProductProvider" android:name="ProductProvider"/>
  </application>

</manifest>
```

Now, let's modify `MainActivity` to exploit this newly registered provider. Modify `MainActivity.kt` to contain the following content:

```kotlin
package com.example.contentproviderexample

import android.content.ContentValues
import android.net.Uri
import android.support.v7.app.AppCompatActivity
import android.os.Bundle
import android.text.TextUtils
import android.view.View
import android.widget.Button
import android.widget.EditText
import android.widget.Toast

class MainActivity : AppCompatActivity(), View.OnClickListener {

  private lateinit var etProductName: EditText
  private lateinit var etProductManufacturer: EditText
  private lateinit var btnAddProduct: Button
  private lateinit var btnShowProduct: Button

  override fun onCreate(savedInstanceState: Bundle?) {
    super.onCreate(savedInstanceState)
    setContentView(R.layout.activity_main)
    bindViews()
    setupInstances()
  }
```

```
private fun bindViews() {
  etProductName = findViewById(R.id.et_product_name)
  etProductManufacturer = findViewById(R.id.et_product_manufacturer)
  btnAddProduct = findViewById(R.id.btn_add_product)
  btnShowProduct = findViewById(R.id.btn_show_products)
}

private fun setupInstances() {
  btnAddProduct.setOnClickListener(this)
  btnShowProduct.setOnClickListener(this)
  supportActionBar?.hide()
}

private fun inputsValid(): Boolean {
  var inputsValid = true
  if (TextUtils.isEmpty(etProductName.text)) {
    etProductName.error = "Field required."
    etProductName.requestFocus()
    inputsValid = false

  } else if (TextUtils.isEmpty(etProductManufacturer.text)) {
    etProductManufacturer.error = "Field required."
    etProductManufacturer.requestFocus()
    inputsValid = false
  }

  return inputsValid
}

private fun addProduct() {
  val contentValues = ContentValues()

  contentValues.put(ProductProvider.NAME, etProductName.text.toString())
  contentValues.put(ProductProvider.MANUFACTURER,
                    etProductManufacturer.text.toString())
  contentResolver.insert(ProductProvider.CONTENT_URI, contentValues)

  showSaveSuccess()
}
```

The following function is called to show products that exist in the database:

```
private fun showProducts() {
  val uri = Uri.parse(ProductProvider.URL)
  val cursor = managedQuery(uri, null, null, null, "name")

  if (cursor != null) {
    if (cursor.moveToFirst()) {
```

```
            do {
               val res = "ID: ${cursor.getString(cursor.getColumnIndex
                        (ProductProvider.ID))}" + ",
               \nPRODUCT NAME: ${cursor.getString(cursor.getColumnIndex
                        ( ProductProvider.NAME))}" + ",
               \nPRODUCT MANUFACTURER: ${cursor.getString(cursor.getColumnIndex
                        (ProductProvider.MANUFACTURER))}"

               Toast.makeText(this, res, Toast.LENGTH_LONG).show()
             } while (cursor.moveToNext())
         }
      } else {
         Toast.makeText(this, "Oops, something went wrong.",
                     Toast.LENGTH_LONG).show()
      }
   }

   private fun showSaveSuccess() {
      Toast.makeText(this, "Product successfully saved.",
                  Toast.LENGTH_LONG).show()
   }

   override fun onClick(view: View) {
      val id = view.id

      if (id == R.id.btn_add_product) {
         if (inputsValid()) {
            addProduct()
         }
      } else if (id == R.id.btn_show_products) {
         showProducts()
      }
   }
}
```

The two methods you should focus your attention on in the preceding code block are
addProduct() and showProducts(). addProduct() stores the product data in a
contentValues instance and then inserts this data into the SQLite database with the help
of the ProductProvider by invoking
contentResolver.insert(ProductProvider.CONTENT_URI,
contentValues). showProducts() uses a Cursor to display the product information
stored in the database in toast messages.

Now that we understand what's going on, let's run the application. Build and run the application as you have done thus far and wait until the application installs and starts. You will be taken straight to `MainActivity` and presented with a form to input the name and manufacturer of a product:

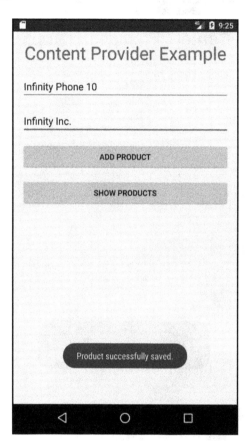

Upon inputting valid product information, click **ADD PRODUCT**. The product will be inserted as a new record in the products table of the application's SQLite database. Add a few more products with the form and click **SHOW PRODUCTS**:

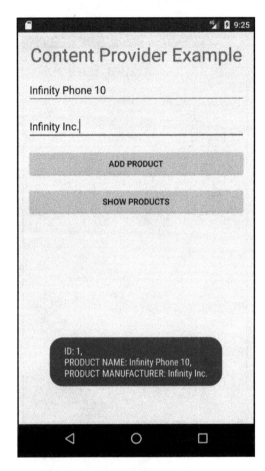

Doing this will lead to the invocation of showProducts() in MainActivity. All product records will be fetched and displayed in toast messages one after the other.

That is as much as we need to implement in a sample application to demonstrate how content providers work. Try to make the application even more awesome by implementing functionality for updating and deleting product records. Doing so will be good practice!

Summary

In this chapter, we dived headfirst into the various mediums of data storage that the Android application framework puts at our disposal. We took a look at how we can use internal storage and external storage to store data in private and public files. In addition, we learned how to work with cache files with the help of both internal and external storage.

Further into the chapter, we learned about the SQLite RDBMS and explored how we can make use of it in our Android applications. We learned how to utilize Room to retrieve and store data in a SQLite database and then went a step further by exploring how to use content providers to control access to data with a SQLite database as an underlying datastore.

In the next chapter, we will conclude our Android application framework exploration by learning how to secure and deploy an Android application.

8

Securing and Deploying an Android App

Over the past six chapters, we have explored the application of Kotlin in the mobile app domain by focusing on Android development. The last chapter explored the various storage mediums provisioned by the Android application framework to the application development. We explored internal storage, external storage, network storage, and SQLite; we also created programs that utilized them. We went a step further by covering the use of Room and content providers to store and retrieve data from a SQLite database.

In this chapter, we willl wrap up our exploration of Android by covering two extremely important topics:

- Android application security
- Android application deployment

We will kick off this chapter by discussing Android application security.

Securing an Android application

It should come as no surprise that security is an important consideration when building software. Besides the security measures put in place in the Android operating system, it is important that developers pay extra attention to ensuring that their applications meet the set security standards. In this section, a number of important security considerations and best practices will be broken down for your understanding. Following these best practices will make your applications less vulnerable to malicious programs that may be installed on a client device.

Data storage

All things being equal, the privacy of data saved by an application to a device is the most common security concern in developing an Android application. Some simple rules can be followed to make your application data more secure.

Using internal storage

As we saw in the previous chapter, internal storage is a good way to save private data on a device. Every Android application has a corresponding internal storage directory in which private files can be created and written to. These files are private to the creating application, and as such cannot be accessed by other applications on the client device. As a rule of thumb, if data should only be accessible by your application and it is reasonably possible to store it in internal storage, do so. Feel free to refer to the previous chapter for a refresher on how to use internal storage.

Using external storage

External storage files are not private to applications, and, as such, can be easily accessed by other applications on the same client device. As a result of this, you should consider encrypting application data before storing it in external storage. There are a number of libraries and packages that can be used to encrypt data prior to its saving to external storage. Facebook's Conceal (`http://facebook.github.io/conceal/`) library is a good option for external-storage data encryption.

In addition to this, as another rule of thumb, do not store sensitive data in external storage. This is because external storage files can be manipulated freely. Validation should also be performed on input retrieved from external storage. This validation should be done as a result of the untrustworthy nature of data stored in external storage.

Using content providers

As you know from the previous chapter, content providers can either prevent or enable external access to your application data. Use the `android:exported` attribute when registering your content provider in the manifest file to specify whether external access to the content provider should be permitted. Set `android:exported` to `true` if you wish the content provider to be exported, otherwise set the attribute to `false`.

In addition to this, content provider query methods—for example, `query()`, `update()`, and `delete()`—should be used to prevent SQL injection (a code injection technique that involves the execution of malicious SQL statements in an entry field by an attacker).

Networking security considerations

There are a number of best practices that should be followed when performing network transactions via an Android application. These best practices can be split into different categories. We shall speak about **Internet Protocol** (**IP**) networking and telephony networking best practices in this section.

IP networking

When communicating with a remote computer via IP, it is important to ensure that your application makes use of HTTPs wherever possible (thus wherever it is supported in the server). One major reason for doing this is because devices often connect to insecure networks, such as public wireless connections. HTTPs ensure encrypted communication between clients and servers, regardless of the network they are connected to. In Java, an `HttpsURLConnection` can be used for secure data transfer over a network. It is important to note that data received via an insecure network connection should not be trusted.

Telephony networking

In instances where data needs to be transferred freely across a server and client applications, **Firebase Cloud Messaging** (**FCM**)—along with IP networking—should be utilized instead of other means, such as the **Short Messaging Service** (**SMS**) protocol. FCM is a multi-platform messaging solution that facilitates the seamless and reliable transfer of messages between applications.

SMS is not a good candidate for transferring data messages, because:

- It is not encrypted
- It is not strongly authenticated
- Messages sent via SMS are subject to spoofing
- SMS messages are subject to interception

Input validation

The validation of user input is extremely important in order to avoid security risks that may arise. One such risk, as explained in the *Using content providers* section, is SQL injection. The malicious injection of SQL script can be prevented by the use of parameterized queries and the extensive sanitation of inputs used in raw SQL queries.

In addition to this, inputs retrieved from external storage must be appropriately validated because external storage is not a trusted data source.

Working with user credentials

The risk of phishing can be alleviated by reducing the requirement of user credential input in an application. Instead of constantly requesting user credentials, consider using an authorization token. Eliminate the need for storing usernames and passwords on the device. Instead, make use of a refreshable authorization token.

Code obfuscation

Before publishing an Android application, it is imperative to utilize a code obfuscation tool, such as **ProGuard**, to prevent individuals from getting unhindered access to your source code by utilizing various means, such as decompilation. ProGuard is prepackaged included within the Android SDK, and, as such, no dependency inclusion is required. It is automatically included in the build process if you specify your build type to be a release. You can find out more about ProGuard here: `https://www.guardsquare.com/en/proguard`.

Securing broadcast receivers

By default, a broadcast receiver component is exported and as a result can be invoked by other applications on the same device. You can control access of applications to your apps's broadcast receiver by applying security permissions to it. Permissions can be set for broadcast receivers in an application's manifest file with the `<receiver>` element.

Dynamically loading code

In scenarios in which the dynamic loading of code by your application is necessary, you must ensure that the code being loaded comes from a trusted source. In addition to this, you must make sure to reduce the risk of tampering code at all costs. Loading and executing code that has been tampered with is a huge security threat. When code is being loaded from a remote server, ensure it is transferred over a secure, encrypted network. Keep in mind that code that is dynamically loaded runs with the same security permissions as your application (the permissions you defined in your application's manifest file).

Securing services

Unlike broadcast receivers, services are not exported by the Android system by default. The default exportation of a service only happens when an intent filter is added to the declaration of a service in the manifest file. The `android:exported` attribute should be used to ensure services are exported only when you want them to be. Set `android:exported` to `true` when you want a service to be exported and `false` otherwise.

Launching and publishing your Android application

So far, we have taken an in-depth look at the Android system, application development in Android, and some other important topics, such as Android application security. It is time for us to cover our final topic for this book pertaining to the Android ecosystem—launching and publishing an Android application.

You may be wondering at this juncture what the words launch and publish mean. A launch is an activity that involves the introduction of a new product to the public (end users). Publishing an Android application is simply the act of making an Android application available to users. Various activities and processes must be carried out to ensure the successful launch of an Android application. There are 15 of these activities in all. They are:

- Understanding the Android developer program policies
- Preparing your Android developer account
- Localization planning
- Planning for simultaneous release
- Testing against the quality guideline

- Building a release-ready APK
- Planning your application's Play Store listing
- Uploading your application package to the alpha or beta channel
- Device compatibility definition
- Pre-launch report assessment
- Pricing and application distribution setup
- Distribution option selection
- In-app products and subscriptions setup
- Determining your application's content rating
- Publishing your application

Wow! That's a long list. Don't fret if you don't understand everything on the list. Let's look at each item in more detail.

Understanding the Android developer program policies

There is a set of developer program policies that were created for the sole purpose of making sure that the Play Store remains a trusted source of software for its users. Consequences exist for the violation of these defined policies. As a result, it is important that you peruse and fully understand these developer policies—their purposes and consequences—before continuing with the process of launching your application.

Preparing your Android developer account

You will need an Android developer account to launch your application on the Play Store. Ensure that you set one up by signing up for a developer account and confirming the accuracy of your account details. If you ever need to sell products on an Android application of yours, you will need to set up a merchant account.

Localization planning

Sometimes, for the purpose of localization, you may have more than one copy of your application, with each localized to a different language. When this is the case, you will need to plan for localization early on and follow the recommended localization checklist for Android developers. You can view this checklist here:
`https://developer.android.com/distribute/best-practices/launch/localization-che cklist.html`.

Planning for simultaneous release

You may want to launch a product on multiple platforms. This has a number of advantages, such as increasing the potential market size of your product, reducing the barrier of access to your product, and maximizing the number of potential installations of your application. Releasing on numerous platforms simultaneously is generally a good idea. If you wish to do this with any product of yours, ensure you plan for this well in advance. In cases where it is not possible to launch an application on multiple platforms at once, ensure you provide a means by which interested potential users can submit their contact details so as to ensure that you can get in touch with them once your product is available on their platform of choice.

Testing against the quality guidelines

Quality guidelines provide testing templates that you can use to confirm that your application meets the fundamental functional and non-functional requirements that are expected by Android users. Ensure that you run your applications through these quality guides before launch. You can access these application quality guides here:
`https://developer.android.com/develop/quality-guidelines/index.html`.

Building a release-ready application package (APK)

A release-ready APK is an Android application that has been packaged with optimizations and then built and signed with a release key. Building a release-ready APK is an important step in the launch of an Android application. Pay extra attention to this step.

Planning your application's Play Store listing

This step involves the collation of all resources necessary for your product's Play Store listing. These resources include, but are not limited to, your application's log, screenshots, descriptions, promotional graphics, and videos, if any. Ensure you include a link to your application's privacy policy along with your application's Play Store listing. It is also important to localize your application's product listing to all languages that your application supports.

Uploading your application package to the alpha or beta channel

As testing is an efficient and battle-tested way of detecting defects in software and improving software quality, it is a good idea to upload your application package to alpha and beta channels to facilitate carrying out alpha and beta software testing on your product. Alpha testing and beta testing are both types of acceptance testing.

Device compatibility definition

This step involves the declaration of Android versions and screen sizes that your application was developed to work on. It is important to be as accurate as possible in this step as defining inaccurate Android versions and screen sizes will invariably lead to users experiencing problems with your application.

Pre-launch report assessment

Pre-launch reports are used to identify issues found after the automatic testing of your application on various Android devices. Pre-launch reports will be delivered to you, if you opt in to them, when you upload an application package to an alpha or beta channel.

Pricing and application distribution setup

First, determine the means by which you want to monetize you application. After determining this, set up your application as either a free install or a paid download. After you have set up the desired pricing of your application, select the countries you wish to distribute you applications to.

Distribution option selection

This step involves the selection of devices and platforms—for example, Android TV and Android Wear—that you wish to distribute your app on. After doing this, the Google Play team will be able to review your application. If your application is approved after its review, Google Play will make it more discoverable.

In-app products and subscriptions setup

If you wish to sell products within your application, you will need to set up your in-app products and subscriptions. Here, you will specify the countries that you can sell into and take care of various monetary-related issues, such as tax considerations. In this step, you will also set up your merchant account.

Determining your application's content rating

It is necessary that you provide an accurate rating for the application you are publishing to the Play Store. This step is mandated by the Android Developer Program Policies for good reason. It aids the appropriate age group you are targeting to discover your application.

Publishing your application

Once you have catered for the necessary steps prior to this, you are ready to publish your application to the production channel of the Play Store. Firstly, you will need to roll out a release. A release allows you to upload the APK files of your application and roll out your application to a specific track. At the end of the release procedure, you can publish your application by clicking **Confirm rollout**.

So, that was all we need to know to publish a new application on the Play Store. In most cases, you will not need to follow all these steps in a linear manner, you will just need to follow a subset of the steps—more specifically, those pertaining to the type of application you wish to publish. As always, the best way to understand a set of steps is through an example. Let's go ahead and publish an application we developed in a previous chapter. We will publish the Messenger application to the Play Store in the following section. You may choose to publish any of the applications we previously developed if you prefer.

The first thing we will do in order to publish the Messenger app to the Play Store is create a Google Play Developer Account. Doing this is simple. We will show you how in the following paragraphs. First and foremost, choose your favorite web browser and navigate to the following URL: `https://play.google.com/apps/publish/signup` .

Once you open the web page, you will be asked to sign in with your Google account. After signing in with an appropriate Google account, you will be required to accept the developer program agreement. This second step is shown in the following screenshot:

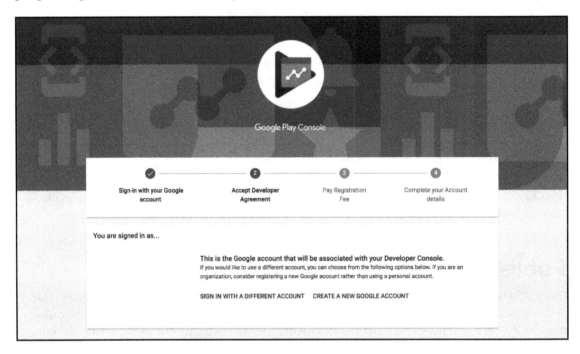

Accept the Google Play Developer agreement by scrolling to the bottom of the web page, as shown in the following screenshot, and ticking **I agree and I am willing to associate my account registration with the Google Play Developer distribution agreement**:

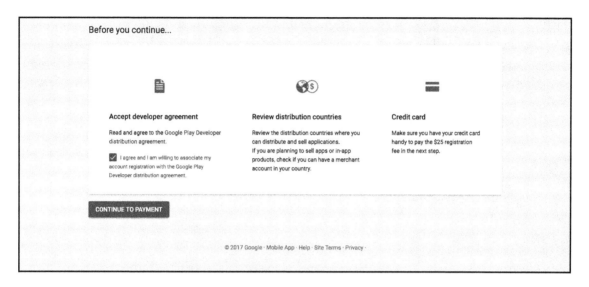

After accepting the agreement, click **CONTINUE TO PAYMENT** to proceed with your developer account creation. You are required to pay a one-time Google Play Developer Account registration fee of $25. You will be taken through a hassle-free payment process. Once payment has been successfully made, you will be prompted accordingly:

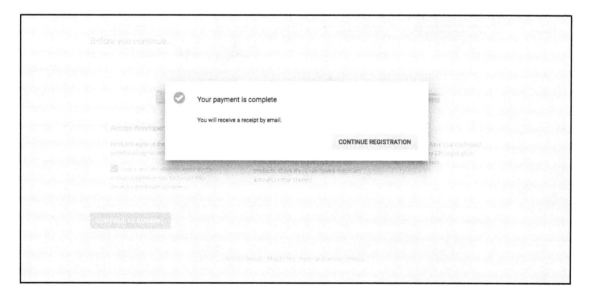

Clicking **CONTINUE REGISTRATION** will take you to the final step of the registration process where you will be asked to complete your account details, as follows:

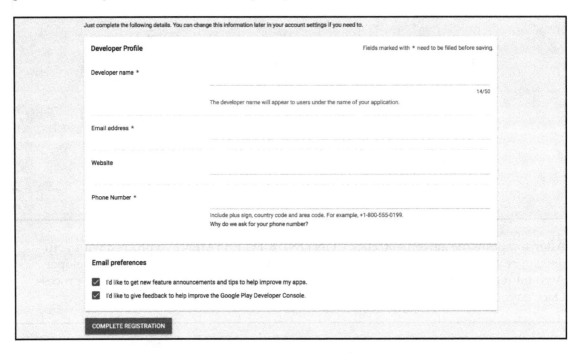

Enter the rest of your account details as required and click **COMPLETE REGISTRATION** to finish the account registration process.

Upon completing registration, you will be directed to the Google Play Developer Console. From here, you can manage your applications, use Google Play game services, manage your orders, download application reports, view alerts, and manage your console settings:

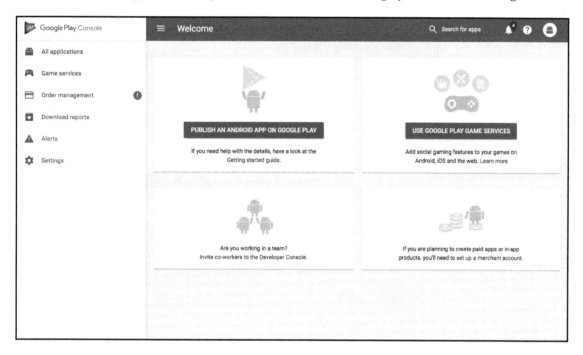

We are interested in publishing an Android application. As such, click **PUBLISH AN ANDROID APP ON GOOGLE PLAY** on the console dashboard. Select a default language and input `Messenger` as the app's title when requested to then click **CREATE**. A new draft application will be created for you on the Developer console:

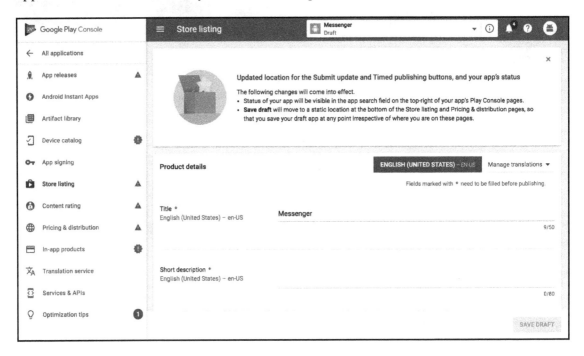

Before proceeding with our application publication process, we must sign a release APK for the Messenger app.

Signing your application for release

Open your Messenger app project in Android Studio. Though not the only method of signing apps, Android Studio will be used in this chapter to sign the Messenger App. First things first, generate a private key for signing by running the following command in your Android Studio terminal:

```
keytool -genkey -v -keystore my-release-key.jks -keyalg RSA -keysize 2048 -
validity 10000 -alias my-alias
```

Running the preceding command prompts you to input a keystore password as well as to provide additional information for your key. The keystore is then generated as a file called my-release-key.jks and saved in the current directory. The key contained in the keystone is valid for 10,000 days.

Now that we have generated a private key, we will configure Gradle to sign our APK. Open the module-level build.gradle file and add a signingConfigs {} block—within the android {} block—with entries for storeFile, storePassword, keyAlias, and keyPassword. Once that has been done, pass that object to the signingConfig property in your app's release build type. Take the following snippet as an example:

```
android {
  compileSdkVersion 26
  buildToolsVersion "26.0.2"
  defaultConfig {
      applicationId "com.example.messenger"
      minSdkVersion 16
      targetSdkVersion 26
      versionCode 1
      versionName "1.0"
      testInstrumentationRunner
"android.support.test.runner.AndroidJUnitRunner"
      vectorDrawables.useSupportLibrary = true
  }

  signingConfigs {
    release {
      storeFile file("../my-release-key.jks")
      storePassword "password"
      keyAlias "my-alias"
      keyPassword "password"
    }
  }
  buildTypes {
    release {
      minifyEnabled false
      proguardFiles getDefaultProguardFile('proguard-android.txt'),
'proguard-rules.pro'
      signingConfig signingConfigs.release
    }
  }
}
```

After doing the preceding, you are ready to sign your APK; before we do this, we must modify our current package name. The `com.example` package name is restricted by Google Play, and, as such, we must change our package name before we attempt to publish our application to the Play Store. Don't fret: changing the name of our app's root package is easy with Android Studio. Firstly, ensure you have set Android Studio to show the project directory structure. This can be done by clicking the dropdown toward the top left of the IDE window and selecting **Project**:

Once you have done the preceding, unhide all empty middle packages in the project structure view by deselecting the **Hide Empty Middle Packages** option in the project structure settings menu, as shown in the following screenshot:

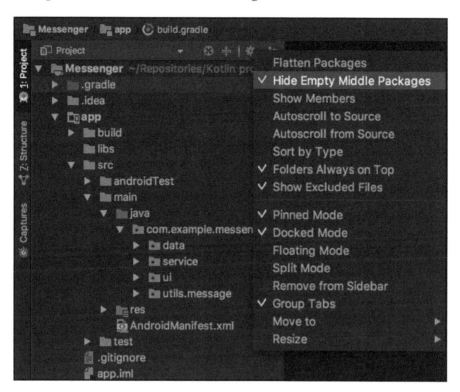

After deselecting the preceding option, empty middle packages will no longer be hidden, and, as such, `com.example.messenger` will be split into three visible packages: `com`, `example`, and `messenger`. Let's rename the `example` package to something else. Change `example` to a name obtained from the combination of your first and last names. So, if your first name and surname are Kevin Fakande, the package name will be renamed from `example` to `kevinfakande`. A package can be renamed by right-clicking on it and selecting **Refactor | Rename...**. After your package has been renamed, check your manifest and `build.gradle` files to ensure that the modification to your project's package if reflected. Thus, wherever you see the `com.example.messenger` string in your `build.gradle` or manifest files, modify it to `com.{full_name}.messenger`.

Having made the preceding changes, you are ready to sign your application. Type the following command in your Android Studio terminal:

```
./gradlew assembleRelease
```

Running the preceding command creates a release APK that has been signed with your private key in the `<project_name>/<module_name>/build/outputs/apk/release` path. The APK will be named as `<module_name>-release.apk`. As our module in this project is named `app`, the APK in this case will be named `app-release.apk`. APKs that have been signed with a private key are ready for distribution. Having signed our APK, we are ready to finish the publication of the Messenger app.

Releasing your Android app

Having signed your Messenger application, you can proceed with completing the required application details toward the goal of releasing your app. Firstly, you need to create a suitable store listing for the application. Open the Messenger app in the Google Play Console and navigate to the store-listing page (this can be done by selecting **Store Listing** on the side navigation bar).

You will need to fill out all the required information in the store listing page before we proceed further. This information includes product details, such as a title, short description, full description, as well as graphic assets and categorization information—including the application type, category and content rating, contact details, and privacy policy. The Google Play Console store listing page is shown in the following screenshot:

Once the store listing information has been filled in, the next thing to fill in is the pricing and distribution information. Select **Pricing & distribution** on the left navigation bar to open up its preference selection page. For the sake of this demonstration, we set the pricing of this app to **FREE**. We also selected five random countries to distribute this application to. These countries are Nigeria, India, the United States of America, the United Kingdom, and Australia:

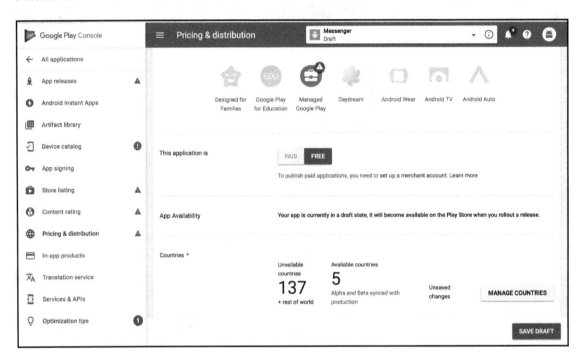

Besides selecting the type of pricing and the available countries for product distribution, you will need to provide additional preference information. The necessary information to be provided includes device category information, user program information, and consent information.

It is now time to add our signed APK to our Google Play Console app. Navigate to **App releases** | **MANAGE BETA** | **EDIT RELEASE**. In the page that is presented to you, you may be asked whether you want to opt into Google play app signing:

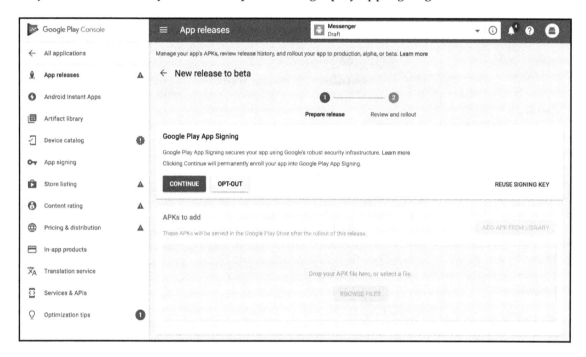

For the sake of this example, select **OPT-OUT**. Once **OPT-OUT** is selected, you will be able to choose your APK file for upload from your computer's file system. Select your APK for upload by clicking **BROWSE FILES**, as shown in the following screenshot:

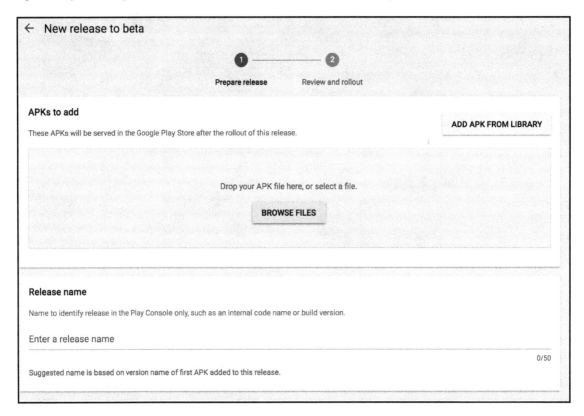

After selecting an appropriate APK, it will be uploaded to the Google Play Console. Once the upload is done, the play console will automatically add a suggested release name for your beta release. This release name is based on the version name of the uploaded APK. Modify the release name if you are not comfortable with the suggestion. Next, add a suitable release note in the text field provided. Once you are satisfied with the data you have input, save and continue by clicking the **Review** button at the bottom of the web page. After reviewing the beta release, you can roll it out if you have added beta testers to your app. Rolling out a beta release is not our focus, so let's divert back to our main goal: publishing the Messenger app.

Having uploaded an APK for your application, you can now complete the mandatory content rating questionnaire. Click the **Content rating** navigation item on the sidebar and follow the instructions to do this. Once the questionnaire is complete, appropriate ratings for your application will be generated:

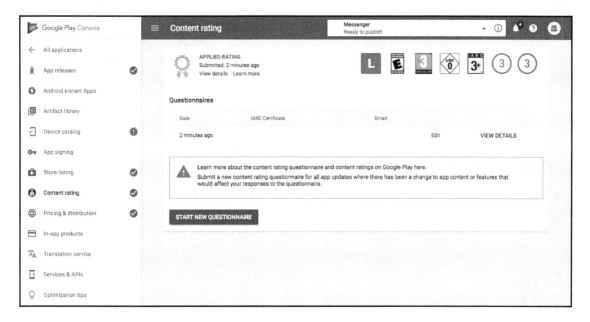

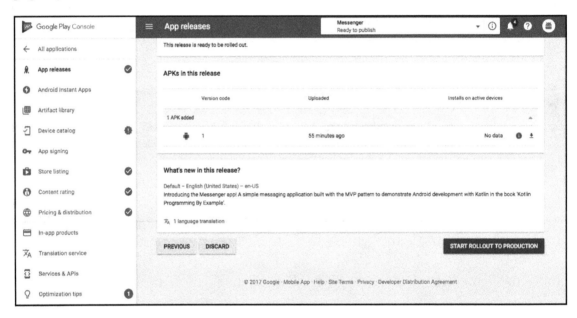

With the completion of the content rating questionnaire, the application is ready to be published to production. Applications that are published to production are made available to all users on the Google Play Store. On the play console, navigate to **App releases** | **Manage Production** | **Create releases**. When prompted to upload an APK, click the **ADD APK FROM LIBRARY** button to the right of the screen and select the APK we previously uploaded (the APK with a version name of 1.0) and complete the necessary release details similar to how we did when creating a beta release. Click the review button at the bottom of the page once you are ready to proceed. You will be given a brief release summary in the page that follows:

Go through the information presented in the summary carefully. Start the roll out to production once you have asserted that you are satisfied with the information presented to you in the summary. Once you start the roll out to production, you will be prompted to confirm your understanding that your app will become available to users of the Play Store:

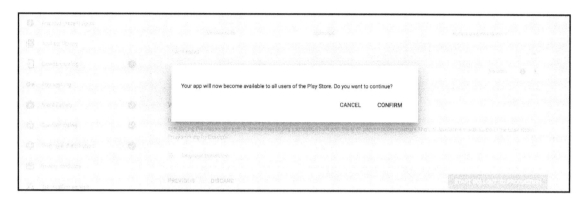

Click **Confirm** once you are ready for the Messenger app to go live on the Play Store. Congratulations! You have now published your first application to the Google Play Store!

Summary

In this chapter, we concluded our exploration of the Android application framework by learning how to secure and publish Android applications to the Google Play Store. We identified security threats to Android applications and fully explained ways to alleviate them, we also noted best practices to follow when developing applications for the Android ecosystem. We learned how to work with storage mediums and carry out networking processes securely. In addition to this, we learned how to secure Android components, such as services and broadcast receivers.

Finally, we took a deep dive into the process of application publication to the Play Store. Besides covering all the necessary steps for the successful publication of an Android application, we went a step further and published the Messenger application to the Google Play Store.

In the next chapter, we will begin the exploration of Kotlin in the development of web applications by developing a place reviewer application.

Creating the Place Reviewer Backend with Spring

9

We focused on the utilization of Kotlin in the development of applications for the Android platform in the previous four chapters. The chapter we just concluded covered, extensively, the various activities involved in securing and deploying an Android application. We took a look at some best practices—in relation to security—when working with data storage as well as when communicating over a network. In addition, we discussed the necessary security considerations when handling user inputs and working with user credentials.

Furthermore, we looked at various ways of securing some Android application components—such as services and broadcast receivers. Lastly, we took a step-by-step approach to properly deploying an Android application to the Google Play Store. In this chapter, we will have an in-depth look at how Kotlin can be used to develop web-based solutions – specifically with Spring—by developing the Place Reviewer application. This chapter will focus on the development of the backend of the Place Reviewer application, and the following chapter will focus on its frontend. In the course of reading through this chapter, you will learn about:

- The Model-View-Controller design pattern
- Logstash and its use in centralizing, transforming, and stashing data.
- Securing a website with Spring Security

Let's dive right in to what we have to learn by first taking a look at the Model-View-Controller design pattern.

The MVC design pattern

The MVC pattern, also known as the Model-View-Controller pattern, is an application design pattern that is used primarily for the separation of concerns within modern applications. More specifically, it is a design pattern for user interfaces that divides an application, primarily, into three distinct components. This separation of application modules into distinct parts is done for several reasons. One such reason is to isolate presentation logic from core business logic. Let us take a look at these three application components in the MVC pattern.

The model

The model is the component that is in charge of the management of data and logic of an MVC application. As the model is the principal manager of all data and business logic, you can view it as the powerhouse of an MVC application.

The view

This is a visual representation of data that exists in and is generated by an application. It is the primary point of interaction that a user has with the application.

The controller

The controller is an intermediary actor between the view and the model. It is in charge of retrieving input – primarily from the view – and feeding an appropriately transformed form of the input to the model. It is also in charge of updating the view with data whenever the need arises:

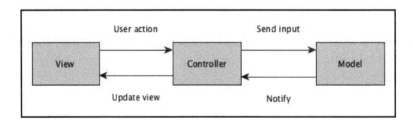

Designing and implementing the Place Reviewer backend

Since we have previously had hands-on experience with the process of designing a system, in this chapter, we are going to focus less on the processes involved in the designing of the Place Reviewer system. Instead, we are going to make a number of quick use case specifications for the system, identify the necessary entities that will be required for the implementation of our system's database, and delve right into the development of the system. Let us go ahead and state our use cases for the Place Reviewer system.

Use case identification

As we did earlier on in this book, we shall commence our use case specification procedure by firstly identifying the actors of the system. Before we can identify the actors of the system, we must have a thorough understanding of what the Place Reviewer web application can do.

As you may have figured out already, the Place Reviewer web application is an internet-based application that facilitates the frictionless creation of location reviews by users of the platform. Once a user has been registered, he/she is able to utilize the platform to create an opinionated review (a review based on personal experiences) of any location in the world. The user will be able to select the location that he/she wants to review with the help of a map.

Now that we understand what the Place Reviewer application can do, we can go ahead and identify the actors in the Place Reviewer system. As you will have surmised by this time, the implementation of the Place Reviewer application we are going to make only has one actor—the user. The use cases of the user are as follows:

- The user uses the Place Reviewer application to create location reviews
- The user uses the Place Reviewer application to view reviews that were created by other users
- The user can view the exact location that was reviewed, by another user, on an interactive map
- The user can register on the Place Reviewer platform
- The user can logout from his/her Place Reviewer account

We have made suitable progress thus far. We have been able to state unequivocally what the Place Reviewer system does, identify the actors of the system, and clearly state the use cases of the system by its sole actor - the user. Let us go a step further by identifying the data that the system will need to cater for.

Identifying data

As a consequence of our previous use case definitions, we can easily identify the type of data that the Place Reviewer application must cater for—by the creation of appropriate models. The first type of data is the user data and the second type is the review data. The user data, as the name implies, is the data relating to a user registered on the platform, whereas the review data is the necessary data for every review created on the platform.

We will require the following data for the user: the user's email address, username, password, and account status. In addition, we will need a unique identifier for each user of the platform—a user ID—and the date on which the user was registered. With respect to the data necessary for reviews, we will need a review title, its body the content of the review, the address of the place reviewed, the name of the place reviewed, positional information of the place reviewed (its longitudinal and latitudinal coordinates), and a place ID to specifically identify the place being reviewed. In addition, a unique identifier will be required for the review being created as well as information relating to the time the review was created.

At this juncture, you might be thinking: hold on, why do we have information pertaining to a place (a place name, place address, place ID, and longitude and latitude) coupled with the information of a review? Why don't we separate this information and consider it a distinct type of data we will be catering for? If you thought this, you are right, that will be a great approach to doing things if we, say, had a database table that possessed all the information of all the places we wanted to be reviewable on the platform. Sadly, we do not have any such table.

Now, you may be wondering: how can we provide a user with the ability to review places of which we have no information about? The answer is simple. We utilize Google's Places API. We shall take a look at how to do this in the next chapter, but for now, hold on tight as we begin implementing the Place Reviewer backend.

Setting up the database

As it is necessary for our system to store information, we need to set up a database for our application to persist data in. As we utilized Postgres as our primary datastore in a previous application we developed, we will utilize it as our primary datastore. We have already covered how to set up Postgres on various systems and as such we will not bother covering that here. Let us go ahead and create our database. Open your Terminal and run the following command:

```
createdb -h localhost —username=<username> place-reviewer
```

Once you run the command, a database named `place-reviewer` will be created on your system. The username you input in place of the `<username>` argument will be the username that you will use to connect to the database. Having set up the database for our application, we can go ahead with the implementation of the backend. We will be utilizing Spring Framework in the implementation of the backend.

Implementing the backend

Having established a sense of direction by specifying the various use cases of our application and setting up a database for our application to connect to, let us go straight ahead with its implementation. Open IntelliJ IDEA and create a new project with the Spring initializer. Upon clicking **Next**, IntelliJ will retrieve the Spring initializer, after which you will be asked to provide certain details for the application. Do the following before proceeding to the next stage of the setup:

1. Input `com.example` as the group ID.
2. Enter `place-reviewer` as the artifact ID.
3. Select **Maven Project** as the project type if it is not already selected.
4. Leave the packaging option and Java version the way they are.
5. Select **Kotlin** as the language. This is important, as we are further learning the Kotlin language, after all.
6. Change the version attribute to 1.0.0.
7. Enter a description of your choice. Ours is `A nifty web application for the creation of location reviews`.
8. Input `com.example.placereviewer` as the package name.

After filling in the required project information, proceed to the next screen by clicking **Next**. We are required to select the dependencies of our project in the screen displayed to us.

 The Spring initializer comes with the Spring plugin, which, at the time of writing, is only available on the IntelliJ IDEA Ultimate Edition, which requires a paid license. If you have the IntelliJ IDEA Community Edition installed, you can still develop this application. Simply generate the project using the Spring initializer utility at `https://start.spring.io` and import the project into IntelliJ IDEA.

Select the Spring **Security**, **Session**, **Cache**, and Web dependencies. In addition, select **Thymeleaf** from the template engine category. Under the SQL category, select **PostgreSQL**. In addition, in the **Spring Boot Version** selection dropdown menu at the top of the screen, select **2.0.0 M7** as the version. Upon selecting the necessary dependencies, the content should be similar to that in the following screenshot:

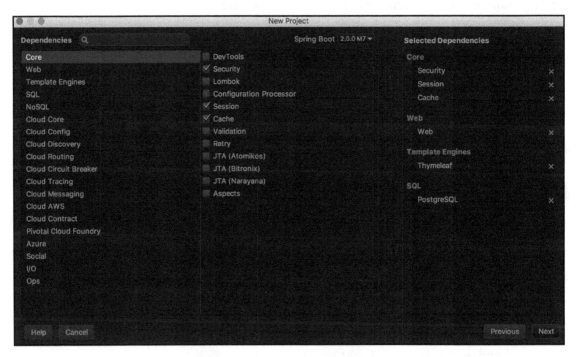

After asserting that you have selected the appropriate dependencies, click **Next** to continue to the final setup screen. Here, you are required to provide a project name and a project location. Fill in `place-reviewer` as the project name and select the location in which you want the project to be saved on your computer:

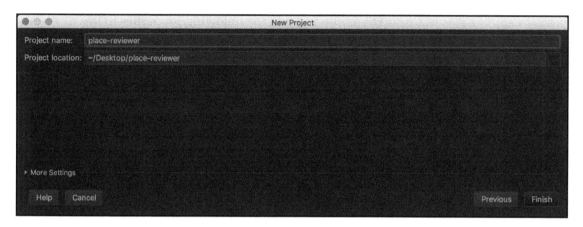

Once this has been done, select **Finish** and wait for the project to be set up. You will be taken to a new IDE window containing the initial project files. We need not give any introductions to the structure of a Spring project, as we have already worked with one in previous chapters. Before we go further, add the following dependencies to the project's pom file:

```
<dependency>
    <groupId>org.springframework.boot</groupId>
    <artifactId>spring-boot-starter-data-jpa</artifactId>
</dependency>
<dependency>
    <groupId>org.webjars</groupId>
    <artifactId>bootstrap</artifactId>
    <version>4.0.0-beta.3</version>
</dependency>
<dependency>
    <groupId>org.webjars</groupId>
    <artifactId>jquery</artifactId>
    <version>3.2.1</version>
</dependency>
```

Now, let's get on with connecting our application to our database.

Connecting the backend to Postgres

To connect the Place Reviewer backend to the PostgreSQL database we created for it, we must modify our project's `application.properties` file to contain the necessary properties that are needed to facilitate a database connection with PostgreSQL. Open the project's `application.properties` file and add the following properties to it:

```
spring.jpa.hibernate.ddl-auto=create-drop
spring.jpa.generate-ddl=true
spring.datasource.url=jdbc:postgresql://localhost:5432/place-reviewer
spring.datasource.driver.class-name=org.postgresql.Driver
spring.datasource.username=<username>
```

Insert an appropriate username where the `<username>` property is within the preceding code snippet. Having added the appropriate database connection properties, Spring Boot will be able to connect to the specified database upon the application start. Having set up the appropriate database connection properties for our project, let us create models for the `User` and `Review` entities we earlier identified.

Creating models

We previously identified two distinct types of entities that must be catered for in our system: the `User` entity and the `Review` entity. It is time to create appropriate models for these entities. The first of these entities we will concern ourselves with is the User. Create a `data` package within the `com.example.placereviewer` package. Add a `model` package within the newly created `data` package. Now, add a `User.kt` file within the newly created `com.example.placereviewer.data.model` package with the following content:

```
package com.example.placereviewer.data.model

import com.example.placereviewer.listener.UserListener
import org.springframework.format.annotation.DateTimeFormat
import java.time.Instant
import java.util.*
import javax.persistence.*
import javax.validation.constraints.Pattern
import javax.validation.constraints.Size


@Entity
@Table(name = "`user`")
@EntityListeners(UserListener::class)
data class User(
```

```
@Column(unique = true)
@Size(min = 2)
@Pattern(regexp = "^[A-Z0-9._%+-]+@[A-Z0-9.-]+\\\\.[A-Z]{2,6}\$")
var email: String = "",
  @Column(unique = true)
var username: String = "",
  @Size(min = 60, max = 60)
var password: String = "",
  @Column(name = "account_status")
@Pattern(regexp = "\\A(activated|deactivated)\\z")
var accountStatus: String = "activated",
  @Id
  @GeneratedValue(strategy = GenerationType.AUTO)
var id: Long = 0,
  @DateTimeFormat
  @Column(name = "created_at")
var createdAt: Date = Date.from(Instant.now())
) {
  @OneToMany(mappedBy = "reviewer", targetEntity =  Review::class)
  private var reviews: Collection<Review>? = null
}
```

We don't need spend time explaining much of what is going on in the preceding code snippet as we have prior experience with the creation of entities in Spring. In the preceding snippet, we defined a User entity with email, username, password, accountStatus, Id, and createdAt properties as its attributes. In addition, we specified that a User has many Review entities. We also specified an entity listener for the entity with the @EntityListener annotation. We have created neither a Review entity nor a UserListener for the User entity. As we are still focused on the User entity, let us focus on creating its entity listener before concerning ourselves with the Review entity. Add a new listener package to com.example.placereviewer and add a UserListener.kt file to it containing the following code:

```
package com.example.placereviewer.listener

import com.example.placereviewer.data.model.User
import org.springframework.security.crypto.bcrypt.BCryptPasswordEncoder
import javax.persistence.PrePersist
import javax.persistence.PreUpdate

class UserListener {

  @PrePersist
  @PreUpdate
  fun hashPassword(user: User) {
    user.password = BCryptPasswordEncoder().encode(user.password)
```

```
    }
}
```

`UserListener` has a single `hashPassword` function, which is invoked before persisting and before updating a `User` entity. The method has the single job of encoding the `password` property of a user into its bcrypt equivalent before persisting it in the database.

Having created the necessary listeners for the `User` entity, let us turn our attention to the definition of a `Review` entity. Create a `Review.kt` file in `com.example.placereviewer.data.models` with the following content:

```kotlin
package com.example.placereviewer.data.model

import org.springframework.format.annotation.DateTimeFormat
import java.time.Instant
import java.util.*
import javax.persistence.*
import javax.validation.constraints.Size

@Entity
@Table(name = "`review`")
data class Review(
  @ManyToOne(optional = false)
  @JoinColumn(name = "user_id", referencedColumnName = "id")
  var reviewer: User? = null,
    @Size(min = 5)
  var title: String = "",
    @Size(min = 10)
  var body: String = "",
    @Column(name = "place_address")
    @Size(min = 2)
  var placeAddress: String = "",
    @Column(name = "place_name")
  var placeName: String = "",
    @Column(name = "place_id")
  var placeId: String = "",
  var latitude: Double = 0.0,
  var longitude: Double = 0.0,
    @Id
    @GeneratedValue(strategy = GenerationType.AUTO)
  var id: Long = 0,
    @DateTimeFormat
    @Column(name = "created_at")
  var createdAt: Date = Date.from(Instant.now())
)
```

As can be seen in the preceding code snippet, we created a `Review` data class with the following properties: `reviewer`, `title`, `body`, `placeAddress`, `placeName`, `placeId`, `latitude`, `longitude`, `id`, and `createdAt`. The reviewer property is of the type `User`. It references the creator of the review. Every review must be created by a user. In addition, many reviews are created by a single user. We use the `@ManyToOne` annotation to properly declare this relationship between the `Review` and `User` entities.

Creating data repositories

As we have now set up our necessary entities, we must create repositories which we will use to access data pertaining to our entities. Create a repositories package within the `com.example.placereviewer` package. We have two entities, and as such, we shall create two repositories (one to access data pertaining to each entity). The first of the repositories will be `UserRepository` and the second will be `ReviewRepository`. Create a `UserRepository` interface file within `com.example.placereviewer.data.repository` with the following content:

```
package com.example.placereviewer.data.repository


import com.example.placereviewer.data.model.User
import org.springframework.data.repository.CrudRepository

interface UserRepository : CrudRepository<User, Long> {

    fun findByUsername(username: String): User?
}
```

The `findByUsername(String)` method retrieves a `User` from the database which has a username that corresponds to that passed as an argument to the function. The following is the `ReviewRepository` interface:

```
package com.example.placereviewer.data.repository

import com.example.placereviewer.data.model.Review
import org.springframework.data.repository.CrudRepository

interface ReviewRepository : CrudRepository<Review, Long> {

    fun findByPlaceId(placeId: String)
}
```

Having set up our entities and repositories to query these entities, we can start work on implementing the core business logic of the Place Reviewer application in the form of services and service implementations.

Place Reviewer business logic implementation

As previously explained, in an application that adheres to the MVC design pattern, there are three primary components of consequence. These components are the model, view, and controller. The models are the components that are in charge of data management and the execution of business logic. In our Place Reviewer application, we are going to implement our models in the form of services that can be used across the backend. At this juncture, we need to create two fundamental services. The first to manage data pertaining to users of the application, and the second to manage review data.

First, we must create a `UserService` interface that defines the behaviors that must be implemented by a valid `UserServiceImpl` class. We previously stated in our use cases for the Place Reviewer application that a user must be able to register (hence create an account) on the platform. As such, we must cater for this process in our model. Create a `service` package in the project's root package. Now, add the `UserService` interface to it:

```
package com.example.placereviewer.service

interface UserService {

    fun register(username: String, email: String, password: String): Boolean
}
```

We declared one method that must be implemented by a valid `UserService`. This method is the `register (String, String, String)` method. `register()` takes three strings as arguments. The first is the username of the user to be registered, the second is a valid email address for the user, and the third is his/her password of choice. When invoked with appropriate arguments, `register()` attempts to register the user with his/her provided credentials and returns `true` if the user was registered successfully; otherwise, it returns `false`.

The following is the implementation of the preceding `UserService`. Add it to the `service` package:

```
package com.example.placereviewer.service

import com.example.placereviewer.data.model.User
import com.example.placereviewer.data.repository.UserRepository
```

```
import org.springframework.stereotype.Service

@Service
class UserServiceImpl(val userRepository: UserRepository) : UserService {

  override fun register(username: String, email: String,
                        password: String): Boolean {
    val user = User(email, username, password)
    userRepository.save(user)

    return true
  }

}
```

The workings of the `register()` function implemented by our `UserServiceImpl` class is straightforward. When valid username, email, and password arguments are passed to it, it creates a new object of the user - passing the appropriate arguments to its constructor. After the creation of an object of the user, the user is saved to the database with the following line:

```
userRepository.save(user)
```

`userRepository` is an instance of the `UserRepository` we created earlier. This instance is injected into the constructor of `UserServiceImpl` automatically by Spring Framework. Once the user has been saved to the database, the Boolean value `true` is returned.

Up next is the implementation of a review service interface. Our review service must facilitate the creation of reviews and the listing of reviews that have been created by users of the platform. As a consequence of this requirement, we will mandate the implementation of the `createReview()` and `listReview()` methods in our user `service` interface.

Add the following `ReviewService` interface to the `service` package of the project:

```
package com.example.placereviewer.service

import com.example.placereviewer.data.model.Review

interface ReviewService {

  fun createReview(reviewerUsername: String, reviewData: Review): Boolean

  fun listReviews(): Iterable<Review>
}
```

The following is the `ReviewServiceImpl` class for the service we have just created. Add it, as well as all the services we will create later on in this chapter, to `com.example.placereviewer.service`:

```
package com.example.placereviewer.service

import com.example.placereviewer.data.model.Review
import com.example.placereviewer.data.model.User
import com.example.placereviewer.data.repository.ReviewRepository
import com.example.placereviewer.data.repository.UserRepository
import org.springframework.stereotype.Service

@Service
class ReviewServiceImpl(val reviewRepository: ReviewRepository, val
userRepository: UserRepository) : ReviewService {

  override fun listReviews(): Iterable<Review> {
    return reviewRepository.findAll()
  }

  override fun createReview(reviewerUsername: String,
                            reviewData: Review): Boolean {
    val reviewer: User? = userRepository.findByUsername(reviewerUsername)

    if (reviewer != null) {
      reviewData.reviewer = reviewer
      reviewRepository.save(reviewData)
      return true
    }

    return false
  }
}
```

`listReviews()` returns an iterable containing all the review data that has been stored within the application's database. `createReview()`, on the other hand, takes a string whose value is the username of the user creating the review and an instance of Review containing the data for the review to be created. `createReview()` first retrieves the user with the specified username by invoking the `findByUsername()` method of `UserRepository`. This user retrieved is the creator of the review—hence, the reviewer.

If a null object is not returned by `UserRepository`, the user exists and as such, the retrieved user is assigned to the `reviewer` property of the review to be saved. After this assignment, the review is saved to the database and the function returns `true` – signifying that the process was successful. If no user with the username provided was found, `false` is returned by `createReview()`.

Having created appropriate models in the form of services, let us work on securing our Place Reviewer application. This is an important procedure as we do not want unauthorized individuals to be able to access our application resources.

Securing the Place Reviewer backend

Similar to how we went about securing the Messenger API in Chapter 4, *Designing and Implementing the Messenger Backend with Spring Boot 2.0*, we shall utilize Spring Security to secure the Place Reviewer backend. Regardless of the utilization of Spring Security here, there is a slight variation to the way we are going to go about securing our application. In Chapter 4, *Designing and Implementing the Messenger Backend with Spring Boot 2.0*, we configured Spring Security to rely explicitly on JSON web tokens for the authorization of client applications. This time, we will rely solely on the power of Spring Security. In doing so, we will not make use of any other technology, such as JSON web tokens. Without further ado, let us begin work on securing our backend.

First and foremost, we must create a custom web security configuration for our application. This custom configuration will implement Spring Framework's `WebSecurityConfigurerAdapter`. Create a `config` package in `com.example.placereviewer` and add the following `WebSecurityConfig` class:

```
package com.example.placereviewer.config

import com.example.placereviewer.service.AppUserDetailsService
import org.springframework.context.annotation.Bean
import org.springframework.context.annotation.Configuration
import org.springframework.http.HttpMethod
import org.springframework.security.authentication.AuthenticationManager
import org.springframework.security.config.BeanIds
import org.springframework.security.config.annotation
        .authentication.builders.AuthenticationManagerBuilder
import org.springframework.security.config.annotation
        .web.builders.HttpSecurity
import org.springframework.security.config.annotation
        .web.configuration.EnableWebSecurity
import org.springframework.security.config.annotation
```

```kotlin
            .web.configuration.WebSecurityConfigurerAdapter
import org.springframework.security.core.userdetails
            .UserDetailsService
import org.springframework.security.crypto.bcrypt
            .BCryptPasswordEncoder
import org.springframework.security.web
            .DefaultRedirectStrategy
import org.springframework.security.web.RedirectStrategy

@Configuration
@EnableWebSecurity
class WebSecurityConfig(val userDetailsService: AppUserDetailsService) :
WebSecurityConfigurerAdapter() {

  private val redirectStrategy: RedirectStrategy =
                              DefaultRedirectStrategy()

  @Throws(Exception::class)
  override fun configure(http: HttpSecurity) {
    http.authorizeRequests()
        .antMatchers(HttpMethod.GET,"/register").permitAll()
        .antMatchers(HttpMethod.POST,"/users/registrations").permitAll()
        .antMatchers(HttpMethod.GET,"/css/**").permitAll()
        .antMatchers(HttpMethod.GET,"/webjars/**").permitAll()
        .anyRequest().authenticated()
        .and()
        .formLogin()
        .loginPage("/login")
        .successHandler { request, response, _ ->
          redirectStrategy.sendRedirect(request, response, "/home")
        }
        .permitAll()
        .and()
        .logout()
        .permitAll()
  }

  @Throws(Exception::class)
  override fun configure(auth: AuthenticationManagerBuilder) {
    auth.userDetailsService<UserDetailsService>(userDetailsService)
        .passwordEncoder(BCryptPasswordEncoder())
  }

  @Bean(name = [BeanIds.AUTHENTICATION_MANAGER])
  override fun authenticationManagerBean(): AuthenticationManager {
    return super.authenticationManagerBean()
  }
}
```

As we have explained earlier in this book, the `configure(HttpSecurity)` method of `WebSecurityConfig` has the task of configuring which HTTP URL paths are to be secured and which are not. With the `configure(HttpSecurity)` method, we have configured Spring Security to permit all HTTP POST requests to `/users/registrations` and GET requests whose paths match the paths `/register`, `/css`, and `/webjars/**`. In addition, we have permitted all HTTP requests to a login page that can be accessed from the path `/login`.

We added a success handler to the login action which utilizes the `redirectStrategy` property that we defined for our `WebSecurityConfig` class to redirect a client to the `/home` path upon successful login of a user. Lastly, we permitted all logout requests to our backend.

`configure(AuthenticationManagerBuilder)` sets up the `UserDetailsService` in use and specifies a password encoder to be used. We made use of a `BcryptPasswordEncoder` in this case. As you may have noticed, we have created no implementation of `UserDetailsService` in our project. Let us do that now. Add `AppUserDetailsService` to the `com.example.placereviewer.service` package:

```
package com.example.placereviewer.service

import com.example.placereviewer.data.repository.UserRepository
import org.springframework.security.core.GrantedAuthority
import org.springframework.security.core.userdetails.User
import org.springframework.security.core.userdetails.UserDetails
import org.springframework.security.core.userdetails.UserDetailsService
import org.springframework.security.core.userdetails
        .UsernameNotFoundException
import org.springframework.stereotype.Service
import java.util.ArrayList

@Service
class AppUserDetailsService(private val userRepository: UserRepository) :
UserDetailsService {

  @Throws(UsernameNotFoundException::class)
  override fun loadUserByUsername(username: String): UserDetails {
    val user = userRepository.findByUsername(username) ?:
        throw UsernameNotFoundException("A user with the username
                                    $username doesn't exist")

    return User(user.username, user.password,
            ArrayList<GrantedAuthority>())
  }
}
```

`loadUsername(String)` attempts to load the `UserDetails` of a user matching the username passed to the function. If a user matching the provided username cannot be found, then a `UsernameNotFoundException` is thrown.

With all this completed, you have successfully set up Spring Security for our backend. Great work!

Now that we have finished work on our entities, repositories, services, service implementations, and Spring Security configuration, we can ideally begin implementing the frontend of our application. However, the implementation of the application's frontend is a task for the next chapter and as such, let us focus on other things. More specifically, let us explore the process of serving web content to a client application with Spring MVC.

Serving web content with Spring MVC

In Spring MVC, HTTP requests are handled by controllers. Controllers are classes that have been annotated with `@Controller`—similar to how we annotate rest controllers with `@RestController`. The best way to understand the way controllers work is to have an example to scrutinize. Let us create a simple Spring MVC controller that handles HTTP GET requests sent to the `/say/hello` path by returning a view, which has the responsibility of rendering an HTML page to a user.

Create a `controller` package in `com.example.placereviewer` and add the following class to it:

```
package com.example.placereviewer.controller

import org.springframework.stereotype.Controller
import org.springframework.web.bind.annotation.GetMapping
import org.springframework.web.bind.annotation.RequestMapping

@Controller
@RequestMapping("/say")
class HelloController {

  @GetMapping("/hello")
  fun hello(): String {
    return "hello"
  }
}
```

As can be seen, the creation of a controller is in no way a complex task. The annotation of `HelloController` with `@Controller` tells Spring that this class is a Spring MVC controller and as such is capable of handling HTTP requests. In addition, annotating `HelloController` with `@RequestMapping("/say")` specifies that the controller handles HTTP requests that have `/say` as their base paths. We defined a `hello()` action within the controller. Since this action was annotated with `@GetMapping("/hello")`, it handles GET requests to the path `/say/hello`. The string returned by `hello()` is the name of the view resource that should be rendered to the client upon the sending of a request to this route.

Since `hello()` requires that a view named `hello` is returned to the client, our next task is to add such a view to our project. Views are generally added to the `templates` folder of a Spring project's `resources` directory. Add a `hello.html` file to the project by right-clicking on the templates and then selecting **New** | **HTML File**:

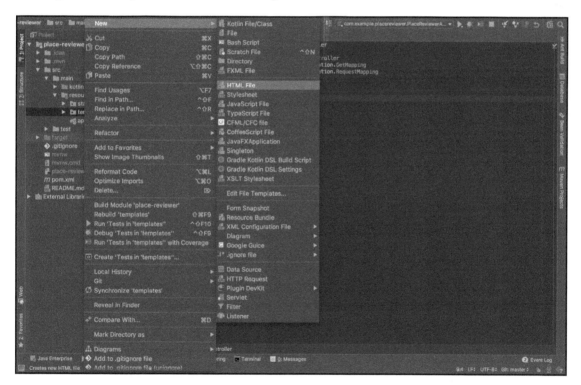

You will be prompted to provide a name for the HTML file to be created. Input `hello` as the name and proceed:

IntelliJ IDEA will generate an HTML file in the selected directory. Once this is done, modify its content to contain the basic HTML, as shown here:

```
<!DOCTYPE html>
<html lang="en">
<head>
  <meta charset="UTF-8">
  <title>Hello</title>
</head>
<body>
Hello world!
</body>
</html>
```

We are now ready to test if the controller we created works. We will know if it works if it returns an HTML page with a message reading `Hello World!` when we send a GET request to its route path. Before we forget, we must add GET requests sent to /say/hello as requests permitted by Spring Security without authentication. Doing this is straightforward; simply modify `configure(HttpSecurity)` in `WebSecurityConfig` to permit GET requests to /say/hello, as shown in the following snippet:

```
@Throws(Exception::class)
override fun configure(http: HttpSecurity) {
    http.authorizeRequests()
        .antMatchers(HttpMethod.GET,"/say/hello").permitAll() // added line
        .antMatchers(HttpMethod.GET,"/register").permitAll()
        .antMatchers(HttpMethod.POST,"/users/registrations").permitAll()
        .antMatchers(HttpMethod.GET, "/css/**").permitAll()
        .antMatchers(HttpMethod.GET, "/webjars/**").permitAll()
```

```
.anyRequest().authenticated()
.and()
.formLogin()
.loginPage("/login")
.successHandler { request, response, _ ->
  redirectStrategy.sendRedirect(request, response, "/home")
}
.permitAll()
.and()
.logout()
.permitAll()
}
```

Build and run the Spring application, and then open your favorite web browser and navigate to the following URL: `http://localhost:5000/say/hello`.

You will be greeted enthusiastically with a `Hello World!` message.

Managing Spring application logs with ELK

When building systems that are intended to be deployed, an important thing to consider is the means by which server log files are managed. A server log is a log file that is created and maintained by a server. Log files generally consist of a list of activities that a server performed. A means of managing application log files that should be strongly considered is the use of the ELK (Elasticsearch, Logstash, and Kibana) stack. In this section, we will learn how to manage Spring application log files with the ELK stack.

Generating logs with Spring

Before getting started with setting up an ELK stack to manage our Spring logs, we must configure Spring to generate log files. This can easily be done with a Spring project's application.properties file. Let's configure our Place Reviewer backend to generate logs.

Open up the project's `application.properties` file and add the following line of code:

```
logging.file=application.log
```

This line of code configures Spring to generate and store server logs in an `application.log` file. This file will be generated and stored in the root directory of your project upon the next start of the project. What we have done is all that's necessary to configure server logs. Now, let's set up our log stack. We shall start by installing Elasticsearch.

Installing Elasticsearch

Elasticsearch can be installed in four easy steps:

1. Download the Elasticsearch packaged in a ZIP file from `https://www.elastic.co/downloads/elasticsearch`.
2. Extract Elasticsearch from the ZIP file upon download.
3. Run Elasticsearch from your terminal. Thus, run `bin/elasticsearch` (`bin/elasticsearch.bat` on Windows):

After executing `bin/elasticsearch` in your terminal, Elasticsearch will run on your system:

```
$ elasticsearch-6.1.1/bin/elasticsearch
[2018-01-16T01:06:00,339][INFO ][o.e.n.Node               ] [] initializing ...
[2018-01-16T01:06:00,461][INFO ][o.e.e.NodeEnvironment    ] [Df8YuN2] using [1]
data paths, mounts [[/ (/dev/disk1)]], net usable_space [11.6gb], net total_spac
e [111.8gb], types [hfs]
[2018-01-16T01:06:00,461][INFO ][o.e.e.NodeEnvironment    ] [Df8YuN2] heap size
[990.7mb], compressed ordinary object pointers [true]
[2018-01-16T01:06:00,483][INFO ][o.e.n.Node               ] node name [Df8YuN2]
derived from node ID [Df8YuN2dQTa5CJH0cSq9KQ]; set [node.name] to override
[2018-01-16T01:06:00,484][INFO ][o.e.n.Node               ] version[6.1.1], pid[
19550], build[bd92e7f/2017-12-17T20:23:25.338Z], OS[Mac OS X/10.12.5/x86_64], JV
M[Oracle Corporation/Java HotSpot(TM) 64-Bit Server VM/1.8.0_65/25.65-b01]
[2018-01-16T01:06:00,484][INFO ][o.e.n.Node               ] JVM arguments [-Xms1
g, -Xmx1g, -XX:+UseConcMarkSweepGC, -XX:CMSInitiatingOccupancyFraction=75, -XX:+
UseCMSInitiatingOccupancyOnly, -XX:+AlwaysPreTouch, -Xss1m, -Djava.awt.headless=
true, -Dfile.encoding=UTF-8, -Djna.nosys=true, -XX:-OmitStackTraceInFastThrow, -
Dio.netty.noUnsafe=true, -Dio.netty.noKeySetOptimization=true, -Dio.netty.recycl
er.maxCapacityPerThread=0, -Dlog4j.shutdownHookEnabled=false, -Dlog4j2.disable.j
mx=true, -XX:+HeapDumpOnOutOfMemoryError, -Des.path.home=/Users/Iyanu/Downloads/
elasticsearch-6.1.1, -Des.path.conf=/Users/Iyanu/Downloads/elasticsearch-6.1.1/c
onfig]
[2018-01-16T01:06:01,607][INFO ][o.e.p.PluginsService     ] [Df8YuN2] loaded mod
ule [aggs-matrix-stats]
```

After running Elasticsearch, you can check if it's functioning properly by running the following command from your terminal:

`curl -XGET http://localhost:9200`

If everything is set up correctly, you will get a response similar to the following:

```
{
  "name" : "Df8YuN2",
  "cluster_name" : "elasticsearch",
  "cluster_uuid" : "Z8SYAKLNSZaMiGkYz7ihfg",
  "version" : {
    "number" : "6.1.1",
    "build_hash" : "bd92e7f",
    "build_date" : "2017-12-17T20:23:25.338Z",
    "build_snapshot" : false,
    "lucene_version" : "7.1.0",
    "minimum_wire_compatibility_version" : "5.6.0",
    "minimum_index_compatibility_version" : "5.0.0"
  },
  "tagline" : "You Know, for Search"
}
```

Installing Kibana

Kibana's installation process is similar to that of Elasticsearch:

1. Download an appropriate Kibana archive from
   `https://www.elastic.co/downloads/kibana`.
2. Extract Kibana from the archive.
3. Run it with `bin/kibana`.

After downloading and running Kibana, check if it works by accessing
`http://localhost:5601/` from your favorite browser. If all is working well, you will be
presented with Kibana's web interface:

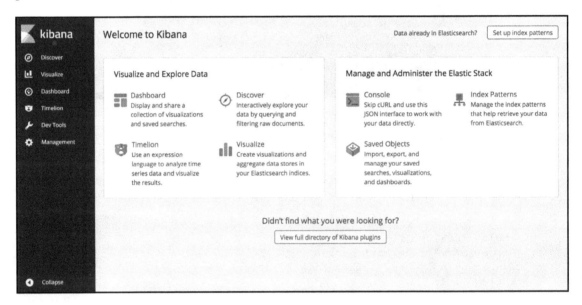

Installing Logstash

To install Logstash:

1. Download its ZIP package from `https://www.elastic.co/downloads/logstash`.
2. Unzip the package.

In the case of Logstash, simply downloading and running it will not suffice. We must configure it to understand the structure of our Spring log file. We do this by creating a Logstash configuration file. A Logstash config file contains three critical sections. These are the input, filter, and output sections. Each section sets up plugins that play a role in the processing of log files. Create a `logstash.conf` file in a suitable directory and add the following code to it:

```
input {
  file {
    type => "java"
    path => "/<path-to-project>/place-reviewer/application.log"
    codec => multiline {
      pattern => "^%{YEAR}-%{MONTHNUM}-%{MONTHDAY} %{TIME}.*"
      negate => "true"
      what => "previous"
    }
  }
}

filter {
  #Tag log lines containing tab character followed by 'at' as stacktrace.
  if [message] =~ "\tat" {
    grok {
      match => ["message", "^(\tat)"]
      add_tag => ["stacktrace"]
    }
  }
  #Grok Spring Boot's default log format
  grok {
    match => [ "message",
               "(?<timestamp>%{YEAR}-%{MONTHNUM}-%{MONTHDAY} %{TIME})
               %{LOGLEVEL:level} %{NUMBER:pid} --- \[(?<thread>
               [A-Za-z0-9-]+)\][A-Za-z0-9.]*\.(?<class>
               [A-Za-z0-9#_]+)\s*:\s+(?<logmessage>.*)",
               "message",
               "(?<timestamp>%{YEAR}-%{MONTHNUM}-%{MONTHDAY} %{TIME})
               %{LOGLEVEL:level} %{NUMBER:pid} --- .+?
               :\s+(?<logmessage>.*)"
             ]
  }

  #Parsing timestamps in timestamp field
  date {
    match => [ "timestamp" , "yyyy-MM-dd HH:mm:ss.SSS" ]
  }
}
```

```
output {
  # Print each event to stdout and enable rubydebug.
  stdout {
    codec => rubydebug
  }
  # Send parsed log events to Elasticsearch
  elasticsearch {
    hosts => ["127.0.0.1"]
  }
}
```

Explaining what all plugins in the preceding code snippet do is beyond the scope of this book. Comments have been added where necessary to facilitate a better understanding. Change `path` in the file plugin of the input section to the absolute path of the Place Reviewer application's `application.log` file.

Once done with the Logstash configuration file, run Logstash with the following command:

```
/bin/logstash -f logstash.conf
```

Logstash should begin storing stashing log events if things were configured properly. The last thing on our agenda is to configure Kibana to read the stashed data.

Configuring Kibana

Kibana can be easily configured to read logs that have been stashed to an Elasticsearch index. Access the Kibana web UI (`http://localhost:5601/`) and navigate to the settings management page by clicking **Management** on the left navigation bar. Our first course of action in configuring Kibana is the creation of an index pattern. Click **Index Patterns** on the management screen to manage the index patterns recognized by Kibana:

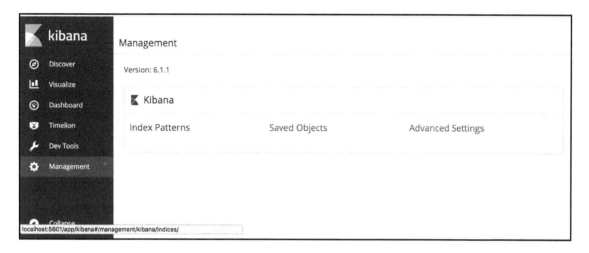

Since you haven't previously created an index pattern on Kibana, you will be prompted to do so:

Input the name of one of the indices recognized by Kibana (displayed on the screen) in the Index pattern field. After inputting an index pattern, proceed to the next step. You will be required to select a time filter field name in the next step:

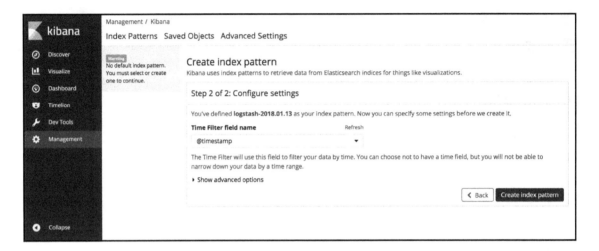

Select @timestamp as the time filter field name in the dropdown. Having selected a time filter field, finish the creation of the index pattern by clicking **Create Index Pattern**. You can manage index patterns you have created at any time from the settings management page by selecting **Index Patterns**. Check your saved patterns:

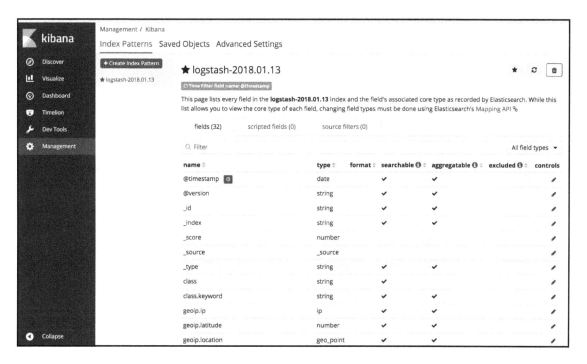

If you see the pattern we just created, then congratulations; you have successfully configured Kibana!

Summary

In this chapter, we went deeper into our exploration of Kotlin and its usefulness in the development of web-based platforms by implementing the backend of the Place Reviewer application. In addition, we learned how to set up a Spring Framework project that utilizes Spring MVC to create modern applications that follow the Model-View-Controller design pattern. Furthermore, we learned how to configure Spring Security to prevent unauthenticated access to Spring web applications. Lastly, we explored the ELK stack and looked at how it can be utilized to manage server logs.

In the next chapter, we will finish building the Place Reviewer application by implementing its frontend. In the process of the frontend implementation, we will learn about how we can build rich web applications with the Google Places API and how to test web applications built with Spring Framework.

10
Implementing the Place Reviewer Frontend

In the previous chapter, we continued our exploration of Kotlin as a viable language for the creation of web applications by commencing the building of the Place Reviewer website. We began the chapter by discussing the Model-View-Controller design pattern and taking a high-level look at the primary components at play in MVC applications: the model, the view, and the controller. Once we had a clear understanding of the MVC design pattern and how it works, we commenced with the design and implementations of the backend for the *Place Reviewer* application.

First and foremost, we identified and clearly stated the anticipated use cases for our application. Next, we identified the data required to build an application that facilitates the use cases identified. After identification of the data to be catered for, we went full steam ahead with the development of the backend. We set up a Postgres database with which our application will communicate, then implemented the necessary entities and models for our application.

Further into the chapter, we discovered how to secure our application—authentication-wise—with Spring Security, this time without the use of JWTs. Finally, we learned how to create controllers for Spring MVC-based applications, as well as how to manage server logs with the ELK stack.

In this chapter, we shall finish up the creation of the *Place Reviewer* application by implementing its frontend. In the process of doing this, we shall learn about the following:

- Working with the Google Places API
- Application testing
- Deploying web applications to AWS

Let's get straight into this chapter by implementing the views for our application.

Creating views with Thymeleaf

As previously defined, a view is a representation of data that exists in and is generated by an application. Views are the primary points of interaction that a user has with an application built with the MVC pattern. The view layer may utilize different technologies to render information to a user. Spring supports a number of view options. These view options are also referred to as templates. Template support in a Spring application is provided by a template engine. Simply put, a template engine enables the utilization of static template files with the view layer of an application. A template engine may also be referred to as a template library. The following template libraries are available for use with Spring:

- Thymeleaf
- JSP/JSTL
- Freemaker
- Tiles
- Velocity

This list is by no means intended to be exhaustive. There are a number of other template libraries available for use with Spring. We shall be relying on Thymeleaf to provide template processing support for our application. If you recall, we included templating support with Thymeleaf upon initial creation of our project. Inspecting the dependencies section of our project's `pom.xml` file will reveal the addition of Thymeleaf to our project:

```
<dependencies>
...
<dependency>
 <groupId>org.springframework.boot</groupId>
 <artifactId>spring-boot-starter-thymeleaf</artifactId>
</dependency>
...
</dependencies>
```

At this juncture, you may desire a more formal definition of what exactly Thymeleaf is. As stated on Thymeleaf's official website, *Thymeleaf is a modern server-side Java template engine for both web and standalone environments. Thymeleaf's main goal is to bring elegant natural templates to your development workflow - HTML that can be correctly displayed in browsers and also work as static prototypes, allowing for stronger collaboration in development teams.* If you require a deeper understanding of Thymeleaf and its goals, you can find more information about it at `http://thymeleaf.org`.

In the previous chapter, we took a simple example of view creation in Spring by implementing a `hello.html` view for the `HelloController`. However, the view we created only displayed a `Hello world!` message to its viewers. We will be creating slightly more complex views in this chapter. We shall start by creating a view facilitating user registration on the platform.

Implementing the user registration view

In this section, we are going to accomplish two tasks. Firstly, we are going to create a view layer that facilitates the registration of new users on the **Place Reviewer** platform. Secondly, we are going to create suitable controllers and actions to present the user with the registration view and handle registration form submissions. Simple enough right? Glad you think so! Go ahead and create a `register.html` template in the `Place Reviewer` project. Recall that all template files belong in the templates directory under resources. Now, add the following template HTML to the file:

```
<!DOCTYPE html>
<html lang="en" xmlns:th="http://www.thymeleaf.org">
<head>
  <title>Register</title>
  <link rel="stylesheet" th:href="@{/css/app.css}"/>
  <link rel="stylesheet"
   href="/webjars/bootstrap/4.0.0-beta.3/css/bootstrap.min.css"/>

  <script src="/webjars/jquery/3.2.1/jquery.min.js"></script>
  <script src="/webjars/bootstrap/4.0.0-beta.3/
   js/bootstrap.min.js"></script>
</head>
<body>
  <nav class="navbar navbar-default nav-enhanced">
    <div class="container-fluid container-nav">
      <div class="navbar-header">
        <div class="navbar-brand">
          Place Reviewer
        </div>
```

```
        </div>
        <ul class="navbar-nav" th:if="${principal != null}">
          <li>
            <form th:action="@{/logout}" method="post">
              <button class="btn btn-danger" type="submit">
                <i class="fa fa-power-off" aria-hidden="true"></i>
                Sign Out
              </button>
            </form>
          </li>
        </ul>
      </div>
    </nav>
    <div class="container-fluid" style="z-index: 2; position: absolute">
      <div class="row mt-5">
        <div class="col-sm-4 col-xs-2">  </div>
        <div class="col-sm-4 col-xs-8">
          <form class="form-group col-sm-12 form-vertical form-app"
           id="form-register" method="post"
           th:action="@{/users/registrations}">
            <div class="col-sm-12 mt-2 lead text-center text-primary">
              Create an account
            </div>
            <hr>
            <input class="form-control" type="text" name="username"
             placeholder="Username" required/>
            <input class="form-control mt-2" type="email" name="email"
             placeholder="Email" required/>
            <input class="form-control mt-2" type="password" name="password"
             placeholder="Password" required/>
            <span th:if="${error != null}" class="mt-2 text-danger"
             style="font-size: 10px" th:text="${error}"></span>
            <button class="btn btn-primary form-control mt-2 mb-3"
             type="submit">
              Sign Up!
            </button>
          </form>
        </div>
        <div class="col-sm-4 col-xs-2"></div>
      </div>
    </div>
    </body>
</html>
```

In this code snippet, we utilized HTML to create a template for the user registration page. The web page in itself is simple. It contains a navigation bar and a form in which a user will input the required registration details for submission. As this is a Thymeleaf template, it should come as no surprise that we utilized some Thymeleaf-specific attributes. Let's take a look at some of these attributes:

- `th:href`: This is an attribute modifier attribute. When it is processed by the templating engine, it computes the link URL to be utilized and sets it in the appropriate tag in which it is used. Examples of tags that this attribute can be used in are `<a>` and `<link>`. We used the `th:href` attribute in the code snippet, as shown here:

  ```
  <link rel="stylesheet" th:href="@{/css/app.css}"/>
  ```

- `th:action`: This attribute works just like the HTML action attribute. It specifies where to send the form data when a form is submitted. The following code snippet specifies that the form data should be sent to an endpoint with the path `/users/registrations`:

  ```
  <form class="form-group col-sm-12 form-vertical form-app"
   id="form-register" method="post"
   th:action="@{/users/registrations}">
   ...
  </form>
  ```

- `th:text`: This attribute is used to specify the text held by a container:

  ```
  <span th:text="Hello world"></span>
  ```

- `th:if`: This attribute can be used to specify whether an HTML tag should be rendered based on the result of a conditional test:

  ```
  <span th:if="${error != null}" class="mt-2 text-danger"
   style="font-size: 10px" th:text="${error}"></span>
  ```

In this code snippet, if a model attribute error exists and its value is not equal to null, then the span tag is rendered on the HTML page; otherwise, it is not rendered.

We also made use of `th:if` in our navigation bar to specify when it should display a button permitting a user to log out of their account:

```
<ul class="navbar-nav" th:if="${principal != null}">
  <li>
    <form th:action="@{/logout}" method="post">
      <button class="btn btn-danger" type="submit">
        <i class="fa fa-power-off" aria-hidden="true"></i> Sign Out
      </button>
    </form>
  </li>
</ul>
```

If the `principal` model attribute is set in the template and it is not `null`, then the sign out button is displayed. The `principal` will always be null unless the user is logged in to their account.

How we added the navigation bar to our template directly may appear to be all right at first glance but it is important we put more thought into what we did. It is not uncommon to make use of a navigation bar DOM element more than once within an application. In fact, this is done very often! We do not want to have to keep rewriting this same code for a navigation bar over and over again in our templates. To avoid this unnecessary repetition, we need to implement the navigation bar as a fragment that can be included at any time within a template.

Create a `fragments` directory within `templates` and add a `navbar.html` file with the following code to it:

```
<!DOCTYPE html>
<html lang="en" xmlns:th="http://www.thymeleaf.org">
  <head>
    <meta charset="UTF-8">
  </head>
  <body>
    <nav class="navbar navbar-default nav-enhanced" th:fragment="navbar">
      <div class="container-fluid container-nav">
        <div class="navbar-header">
          <div class="navbar-brand">
            Place Reviewer
          </div>
        </div>
        <ul class="navbar-nav" th:if="${principal != null}">
          <li>
            <form th:action="@{/logout}" method="post">
              <button class="btn btn-danger" type="submit">
                <i class="fa fa-power-off" aria-hidden="true"></i> Sign Out
```

```
        </button>
      </form>
    </li>
  </ul>
</div>
</nav>
</body>
</html>
```

In this code snippet, we defined a navigation bar fragment available for inclusion in templates with the th:fragment attribute. A defined fragment can be inserted at any time within a template with the use of th:insert. Modify the inner HTML of the <body> tag in register.html to make use of the newly defined fragment as follows:

```
<!DOCTYPE html>
<html lang="en" xmlns:th="http://www.thymeleaf.org">
  <head>
    <title>Register</title>
    <link rel="stylesheet" th:href="@{/css/app.css}"/>
    <link rel="stylesheet"
     href="/webjars/bootstrap/4.0.0-beta.3/css/bootstrap.min.css"/>
    <script src="/webjars/jquery/3.2.1/jquery.min.js"></script>
    <script src="/webjars/bootstrap/4.0.0-beta.3/
     js/bootstrap.min.js"></script>
  </head>
  <body>
    <div th:insert="fragments/navbar :: navbar"></div>
    <!-- inserting navbar fragment -->
    <div class="container-fluid" style="z-index: 2; position: absolute">
      <div class="row mt-5">
        <div class="col-sm-4 col-xs-2">
        </div>
        <div class="col-sm-4 col-xs-8">
          <form class="form-group col-sm-12 form-vertical form-app"
            id="form-register" method="post"
            th:action="@{/users/registrations}">
            <div class="col-sm-12 mt-2 lead text-center text-primary">
              Create an account
            </div>
            <hr>
            <input class="form-control" type="text" name="username"
             placeholder="Username" required/>
            <input class="form-control mt-2" type="email" name="email"
             placeholder="Email" required/>
            <input class="form-control mt-2" type="password"
             name="password" placeholder="Password" required/>
            <span th:if="${error != null}" class="mt-2 text-danger"
```

```
                style="font-size: 10px" th:text="${error}"></span>
                <button class="btn btn-primary form-control mt-2 mb-3"
                type="submit">
                  Sign Up!
                </button>
              </form>
            </div>
            <div class="col-sm-4 col-xs-2"></div>
          </div>
        </div>
    </body>
  </html>
```

As can already be seen, the separation of our navigation bar HTML into a fragment has made our code more succinct and will contribute positively to the quality of our developed templates.

Having created the necessary template for the user registration page, we need to create a controller that will render this template to a visitor of the site. Let's create an application controller. Its job will be to render the web pages of the *Place Reviewer* application to a user upon request.

Add the `ApplicationController` class, shown here, to the controller package:

```
package com.example.placereviewer.controller

import org.springframework.stereotype.Controller
import org.springframework.web.bind.annotation.GetMapping

@Controller
class ApplicationController {

  @GetMapping("/register")
  fun register(): String {
    return "register"
  }
}
```

Nothing special is being done in the code snippet here. We created an MVC controller with a single action that handles a HTTP GET request to the /register path by rendering the register.html view to the user.

We are almost ready to view our newly created registration page. Before we check it out, we must add the `app.css` file required by `register.html`. Static resources such as CSS files should be added to the `static` directory within the application `resource` directory. Add a `css` directory within the `static` directory and add an `app.css` file containing the code shown here to it:

```
//app.css
.nav-enhanced {
  background-color: #00BFFF;
  border-color: blueviolet;
  box-shadow: 0 0 3px black;
}

.container-nav {
  height: 10%;
  width: 100%;
  margin-bottom: 0;
}

.form-app {
  background-color: white;
  box-shadow: 0 0 1px black;
  margin-top: 50px !important;
  padding: 10px 0;
}
```

Great work! Now, go ahead and run the *Place Reviewer* application. Upon starting the app, open your favorite browser and access the web page residing at `http://localhost:5000/register`.

Now, we must implement the logic involved in registering the user. To do this, we must declare an action that accepts the form data sent by the registration form and appropriately processes the data, with the goal of registering the user successfully on the platform. If you recall, we specified that the form data should be sent via POST to `/users/registrations`. Consequently, we need an action that handles such a HTTP request. Add a `UserController` class to the `com.example.placereviewer.controller` package with the code shown here:

```
package com.example.placereviewer.controller

import com.example.placereviewer.component.UserValidator
import com.example.placereviewer.data.model.User
import com.example.placereviewer.service.SecurityService
import com.example.placereviewer.service.UserService
import org.springframework.stereotype.Controller
```

```
import org.springframework.ui.Model
import org.springframework.validation.BindingResult
import org.springframework.web.bind.annotation.GetMapping
import org.springframework.web.bind.annotation.ModelAttribute
import org.springframework.web.bind.annotation.PostMapping
import org.springframework.web.bind.annotation.RequestMapping

@Controller
@RequestMapping("/users")
class UserController(val userValidator: UserValidator,
        val userService: UserService, val securityService: SecurityService) {

  @PostMapping("/registrations")
  fun create(@ModelAttribute form: User, bindingResult: BindingResult,
            model: Model): String {
    userValidator.validate(form, bindingResult)

    if (bindingResult.hasErrors()) {
      model.addAttribute("error", bindingResult.allErrors.first()
                                      .defaultMessage)
      model.addAttribute("username", form.username)
      model.addAttribute("email", form.email)
      model.addAttribute("password", form.password)

      return "register"
    }

    userService.register(form.username, form.email, form.password)
    securityService.autoLogin(form.username, form.password)

    return "redirect:/home"
  }
}
```

create() handles HTTP POST requests sent to /users/registrations. It takes three arguments. The first is form, which is an object of the User class. @ModelAttribute is used to annotate form. @ModelAttribute indicates that the argument should be retrieved by the model. The form model attribute is populated by data submitted by the form to the endpoint. The username, email, and password parameters are all submitted by the registration form. All objects of type User have username, email, and password properties, hence the data submitted by the form is assigned to the corresponding model properties.

The second argument of the function is an instance of `BindingResult`. `BindingResult` serves as a result holder for `DataBinder`. In this case, we used it to bind results of the validation process done by a `UserValidator`, which we are going to create in a bit. The third argument is a `Model`. We use this to add attributes to our model for subsequent access by the view layer.

Before proceeding further with explanations pertaining to the logic implemented in the `create()` action, we must implement both `UserValidator` and `SecurityService`. `UserValidator` has the sole task of validating user information submitted to the backend. Create a `com.example.placereviewer.component` package and include the `UserValidator` class here to it:

```
package com.example.placereviewer.component

import com.example.placereviewer.data.model.User
import com.example.placereviewer.data.repository.UserRepository
import org.springframework.stereotype.Component
import org.springframework.validation.Errors
import org.springframework.validation.ValidationUtils
import org.springframework.validation.Validator

@Component
class UserValidator(private val userRepository: UserRepository) : Validator
{

  override fun supports(aClass: Class<*>?): Boolean {
    return User::class == aClass
  }

  override fun validate(obj: Any?, errors: Errors) {
    val user: User = obj as User
```

Validating that submitted user parameters are not empty. An empty parameter is rejected with an error code and error message:

```
ValidationUtils.rejectIfEmptyOrWhitespace(errors, "username",
        "Empty.userForm.username", "Username cannot be empty")
ValidationUtils.rejectIfEmptyOrWhitespace(errors, "password",
        "Empty.userForm.password", "Password cannot be empty")
ValidationUtils.rejectIfEmptyOrWhitespace(errors, "email",
        "Empty.userForm.email", "Email cannot be empty")
```

Validating the length of a submitted username. A username whose length is less than 6 is rejected:

```
if (user.username.length < 6) {
    errors.rejectValue("username", "Length.userForm.username",
                "Username must be at least 6 characters in length")
}
```

Validating the submitted username does not already exist. A username already taken by a user is rejected:

```
if (userRepository.findByUsername(user.username) != null) {
    errors.rejectValue("username", "Duplicate.userForm.username",
                "Username unavailable")
}
```

Validating the length of a submitted password. Passwords less than 8 characters in length are rejected:

```
if (user.password.length < 8) {
    errors.rejectValue("password", "Length.userForm.password",
                "Password must be at least 8 characters in length")
        }
    }
}
```

UserValidator implements the Validator interface, which is used to validate objects. As such, it overrides two methods: supports(Class<*>?) and validate(Any?, Errors). supports() is used to assert that the validator can validate the object supplied to it. In the case of UserValidator, supports() asserts that the supplied object is an instance of the User class. Hence, all objects of type User are supported for validation by UserValidator.

validate() validates the provided objects. In cases where validation rejections occur, it registers the error with the provided Error object. Ensure you read through the comments placed within the body of the validate() method to get a better grasp of what is going on within the method.

Now, we shall work on `SecurityService`. We will implement a `SecurityService` to facilitate the identification of the currently logged in user and the automatic login of a user after their registration.

Add the `SecurityService` interface here to `com.example.placereviewer.service`:

```
package com.example.placereviewer.service

interface SecurityService {
  fun findLoggedInUser(): String?
  fun autoLogin(username: String, password: String)
}
```

Now, add a `SecurityServiceImpl` class to `com.example.placereviewer.service`. As the name suggests, `SecurityServiceImpl` implements `SecurityService`:

```
package com.example.placereviewer.service

import org.springframework.beans.factory.annotation.Autowired
import org.springframework.security.authentication.AuthenticationManager
import org.springframework.security.authentication.UsernamePasswordAuthenticationToken
import org.springframework.security.core.context.SecurityContextHolder
import org.springframework.security.core.userdetails.UserDetails
import org.springframework.stereotype.Service


@Service
class SecurityServiceImpl(private val userDetailsService:
AppUserDetailsService)
      : SecurityService {

  @Autowired
  lateinit var authManager: AuthenticationManager

  override fun findLoggedInUser(): String? {
    val userDetails = SecurityContextHolder.getContext()
                                    .authentication.details

    if (userDetails is UserDetails) {
      return userDetails.username
    }

    return null
  }
```

```
override fun autoLogin(username: String, password: String) {
  val userDetails: UserDetails = userDetailsService
                                   .loadUserByUsername(username)

  val usernamePasswordAuthenticationToken =
          UsernamePasswordAuthenticationToken(userDetails, password,
                            userDetails.authorities)

  authManager.authenticate(usernamePasswordAuthenticationToken)

  if (usernamePasswordAuthenticationToken.isAuthenticated) {
    SecurityContextHolder.getContext().authentication =
            usernamePasswordAuthenticationToken
  }
 }
}
```

`findLoggedInUser()` returns the username of the currently logged in user. `Username` retrieval is done with the help of Spring Framework's `SecurityContextHolder` class. An instance of `UserDetails` is retrieved by accessing the logged in user's authentication details with a call to `SecurityContextHolder.getContext().authentication.details`. It is important to note that `SecurityContextHolder.getContext().authentication.details` returns an `Object` and not an instance of `UserDetails`. As such, we must do a type check to assert that the object retrieved conforms to the type `UserDetails` as well. If it does, we return the username of the currently logged in user. Otherwise, we return null.

The `autoLogin()` method will be used for the simple task of authenticating a user after registering on the platform. The submitted username and password of the user are passed as arguments to `autoLogin()`, after which an instance of `UsernamePasswordAuthenticationToken` is created for the registered user. Once an instance of `UsernamePasswordAuthenticationToken` is created, we utilize `AuthenticationManager` to authenticate the user's token. If `UsernamePasswordAuthenticationToken` is successfully authenticated, we set the authentication property of the current user to `UsernamePasswordAuthenticationToken`.

Having made our necessary class additions, let's return to our `UserController` to finish up our explanation of the create action. Within `create()`, first and foremost the submitted form input is validated with an instance of `UserValidator`. Errors arising during the course of form data validation are all bound to the instance of `BindingResult` injected into our controller by Spring. Consider these lines of code:

```
if (bindingResult.hasErrors()) {
  model.addAttribute("error", bindingResult.allErrors
                                     .first().defaultMessage)
  model.addAttribute("username", form.username)
  model.addAttribute("email", form.email)
  model.addAttribute("password", form.password)

  return "register"
}
```

`bindingResult` is first checked to assert whether any errors occurred during form data validation. If errors occurred, we retrieve the message of the first error detected and set a `model` attribute error to hold the error message for later access by the view. In addition, we create `model` attributes to hold each input submitted by the user. Lastly, we re-render the registration view to the user.

Notice how we made multiple method invocations for the same `Model` instance in the previous code snippet. There is a much cleaner way we can do this. This involves the use of Kotlin's `with` function:

```
if (bindingResult.hasErrors()) {
  with (model) {
    addAttribute("error", bindingResult.allErrors.first().defaultMessage)
    addAttribute("error", form.username)
    addAttribute("email", form.email)
    addAttribute("password", form.password)
  }
  return "register"
}
```

See how easy and convenient the function is to use? Go ahead and modify `UserController` to make use of `with`, as shown in the preceding code.

You may be wondering why we decided to store a user's submitted data in model attributes. We did this to have a way to reset the data contained in the registration form to what was originally submitted after the re-rendering of the registration view. It will certainly be frustrating for a user to have to input all form data over and over, even if only one form input entered is invalid.

When no input submitted by the user is invalid, the following code runs:

```
userService.register(form.username, form.email, form.password)
securityService.autoLogin(form.username, form.password)
return "redirect:/home"
```

As expected, when the data submitted by a user is valid, he is registered on the platform and logged in to his account automatically. Lastly, he is redirected to his home page. Before we try out our registration form, we must do two things:

- Utilize the model attributes specified in `register.html`
- Create a `home.html` template and a controller to render the template

Luckily for us, both are rather simple to do. First, to utilize the `model` attributes. Modify the form contained in `register.html` as follows:

```
<form class="form-group col-sm-12 form-vertical form-app"
 id="form-register" method="post" th:action="@{/users/registrations}">
  <div class="col-sm-12 mt-2 lead text-center text-primary">
    Create an account
  </div>
  <hr>
  <!-- utilized model attributes with th:value -->
  <input class="form-control" type="text" name="username"
   placeholder="Username" th:value="${username}" required/>
  <input class="form-control mt-2" type="email" name="email"
   placeholder="Email" th:value="${email}" required/>
  <input class="form-control mt-2" type="password" name="password"
   placeholder="Password" th:value="${password}" required/>
  <span th:if="${error != null}" class="mt-2 text-danger"
   style="font-size: 10px" th:text="${error}"></span>
  <button class="btn btn-primary form-control mt-2 mb-3" type="submit">
    Sign Up!
  </button>
</form>
```

Can you spot the changes we made? If you said we used Thymeleaf's `th:value` template attribute to preset the value held by form inputs to their respective `model` attribute values, you are right. Now, let's make a simple `home.html` template. Add the `home.html` template here to the `templates` directory:

```
<html>
  <head>
    <title> Home</title>
  </head>
  <body>
```

```
    You have been successfully registered and are now in your home page.
  </body>
</html>
```

Now, update `ApplicationController` to include an action handling GET requests to
`/home`, as shown here:

```
package com.example.placereviewer.controller

import com.example.placereviewer.service.ReviewService
import org.springframework.stereotype.Controller
import org.springframework.ui.Model
import org.springframework.web.bind.annotation.GetMapping
import java.security.Principal
import javax.servlet.http.HttpServletRequest

@Controller
class ApplicationController(val reviewService: ReviewService) {

  @GetMapping("/register")
  fun register(): String {
    return "register"
  }

  @GetMapping("/home")
  fun home(request: HttpServletRequest, model: Model,
           principal: Principal): String {
    val reviews = reviewService.listReviews()

    model.addAttribute("reviews", reviews)
    model.addAttribute("principal", principal)

    return "home"
  }
}
```

The `home` action retrieves a list of all reviews stored within the database. In addition, the
home action sets a model attribute that holds a principal containing information on the
currently logged in user. Lastly, the home action renders the home page to the user.

Having done what's necessary, let's register a user on the **Place Reviewer** platform. Build and run the application, and access the registration page from your browser (`http://localhost:5000/register`). Firstly, we want to check whether our form validations work by inputting and submitting invalid form data.

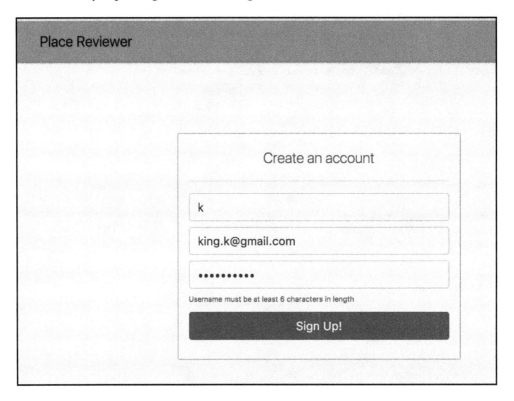

As can be seen, the error was detected by `UserValidator` and was successfully bound to `BindingResult`, then rendered appropriately as an error in the view. Feel free to enter invalid data for other form inputs and ensure the other validations we implemented work as expected. Now to verify that our registration logic works. Input `king.kevin`, `king.k@gmail.com`, and `Kingsman406` in the username, email, and password fields, then click **Sign Up!** A new account will be created and you will be presented with the home page:

It will come as no surprise to you that we are going to make serious modifications to the home page over the course of this chapter. However, for now let's turn our attention towards creating a suitable user login page.

Implementing the login view

Similar to how we went about implementing the user registration view, first and foremost we must work on the view template. The template required for the login view must possess a form that takes the username and password of the user to be logged in as input. We must also provide a button that facilitates the submission of the login form—after all, there is no point having a form if it cannot be submitted. In addition, we must have a means of alerting the user if something goes wrong with the login process, such as in a scenario where the user enters an invalid username and password combination. Lastly, we should provide a link to the account registration page for situations where the viewer of the login page does not already possess an account.

Having identified what is needed from the template to be implemented, let's go ahead with creating it. Add a `login.html` file to the template directory. Now, it is time to work on the template. As always, we must first include the necessary stylesheets and scripts to the template. This is done in the following code snippet:

```
<!DOCTYPE html>
<html lang="en" xmlns:th="http://www.thymeleaf.org">
  <head>
    <title>Login</title>
    <link rel="stylesheet" th:href="@{/css/app.css}"/>
    <link rel="stylesheet"
      href="/webjars/bootstrap/4.0.0-beta.3/css/bootstrap.min.css"/>

    <script src="/webjars/jquery/3.2.1/jquery.min.js"></script>
    <script src="/webjars/bootstrap/4.0.0-beta.3/
      js/bootstrap.min.js"></script>
  </head>
  <body>
  </body>
</html>
```

Having added the style and JavaScript includes necessary for the template, we can now work on the <body> of the template. As we have said before, the body of an HTML template contains the DOM elements that will be rendered to the user upon page load. Add the following code within the <body> tag of login.html:

```
<div th:insert="fragments/navbar :: navbar"></div>

<div class="container-fluid" style="z-index: 2; position: absolute">
  <div class="row mt-5">
    <div class="col-sm-4 col-xs-2"></div>
    <div class="col-sm-4 col-xs-8">
      <form class="form-group col-sm-12 form-vertical form-app"
        id="form-login" method="post" th:action="@{/login}">
        <div class="col-sm-12 mt-2 lead text-center text-primary">
          Login to your account
        </div>
        <hr>
        <input class="form-control" type="text" name="username"
         placeholder="Username" required/>
        <input class="form-control mt-2" type="password"
         name="password" placeholder="Password" required/>
        <span th:if="${param.error}" class="mt-2 text-danger"
         style="font-size: 10px">
          Invalid username and password combination
        </span>
        <button class="btn btn-primary form-control mt-2 mb-3"
         type="submit">
          Go!
        </button>
      </form>
      <div class="col-sm-12 text-center" style="font-size: 12px">
        Don't an account? Register <a href="/register">here</a>
      </div>
    </div>
    <div class="col-sm-4 col-xs-2">
      <div th:if="${param.logout}"
       class="col-sm-12 text-success text-right">
        You have been logged out.
      </div>
    </div>
  </div>
</div>
```

With the body added, the HTML we have created sufficiently describes the required structure of our login page. Along with adding the required form, we added the navigation bar fragment created earlier to the page—no need to write boilerplate code. We also added a means by which a user can be provided feedback pertaining to errors that may arise in the login process. This was done with the following lines:

```
<span th:if="${param.error}" class="mt-2 text-danger"
 style="font-size: 10px">
   Invalid username and password combination
</span>
```

When `param.error` is set, it signifies that an error has occurred during user login, so an **Invalid username and password combination** message is shown to the user. Something to keep in mind is that besides the login page often being the first point of contact a user has with a web application, it can also be a user's last point of contact with the app during an interaction session. This is particularly true in the case of user logouts. After a user is done interacting with an application and logs out, they should be redirected to a login screen. As a result of such a possibility, we added some text to notify a user that they've been logged out:

```
<div th:if="${param.logout}" class="col-sm-12 text-success text-right">
   You have been logged out.
</div>
```

The `<div>` tag is displayed to a user after a successful logout from their account. At this point, ideally we should implement a controller to render `login.html` but if you recall, we have already done so with the use of a custom Spring MVC configuration via our implemented `MvcConfig` class, specifically in the section of code shown here:

```
...

override fun addViewControllers(registry: ViewControllerRegistry?) {
   registry?.addViewController("/login")?.setViewName("login")
}

...
```

We made use of a `ViewControllerRegistry` instance to add a view controller that handles requests to `/login` and set the view to be used to the just-implemented login template. Build and run the application to view the newly implemented view. The web page can be accessed via `http:localhost:5000/login`:

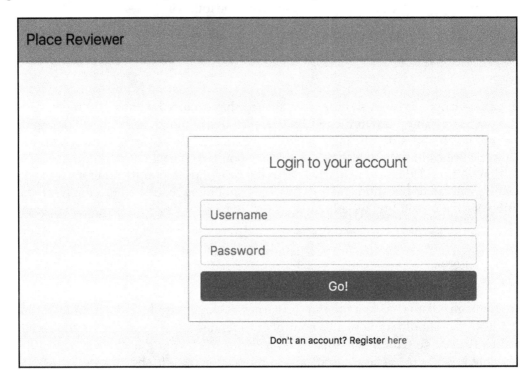

Trying to log in with invalid user credentials will present us with a nice error message:

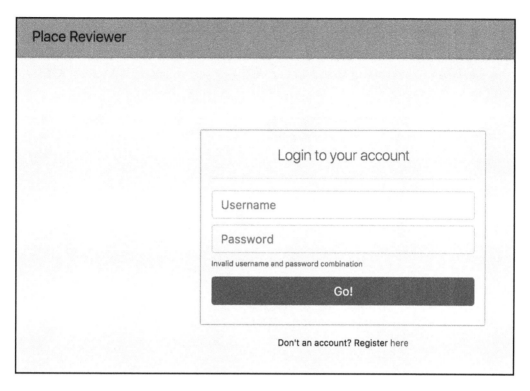

On the other hand, attempting to log in with valid credentials takes us to the application's home page. Speaking of the home page, we need to work on completing its view layer. We are going to need to work directly with the Google Places API from this point onward. As such, we must set up our application to do so before proceeding.

Setting up the Place Reviewer app with the Google Places API web service

The process of setting up a web application with the Google Places API is quick and painless and can be completed in well under five minutes. All in all, the setup can be done in two easy steps:

1. Get an API key
2. Include the Google Places API in your web application

Getting an API key

An API key can be gotten by visiting `https://developers.google.com/places/web-service/get-api-key`, scrolling to the **Get an API key** section, and clicking the **GET A KEY** button:

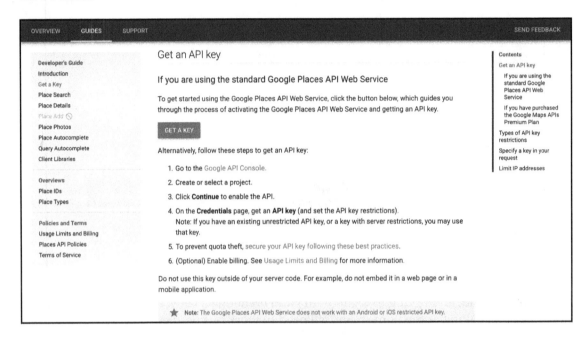

Upon clicking the button, you will be presented with a modal from which you can select or create a project to be integrated with the Google Places API Web Service. Click on the dropdown and select **Create a new project**. You will be requested to provide a project name. Input `Place Reviewer` as the project name:

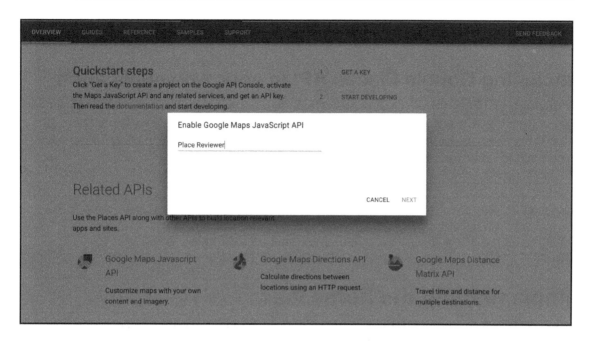

After providing a project name, click **NEXT** to proceed. Your project will be set up for use with the API and you will be presented with an API key to use!

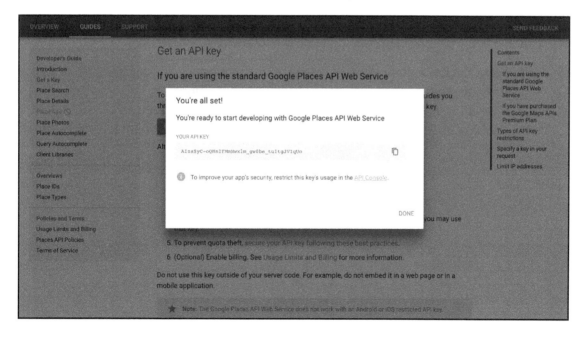

Now that we have an API key, let's have a look at how to use it.

Including Google Places API in your web application

Utilizing an API key for the Google Places API Web Service is just as easy—if not easier—as generating the API key. To make use of a generated API key in your web application, all that is necessary is the inclusion of the following line of HTML in the markup of the page you want to make use of the web service:

```
<script type="text/javascript"
src="https://maps.googleapis.com/maps/api/js?key={{API_KEY}}&libraries=plac
es"></script>
```

Ensure you replace `{{API_KEY}}` with your generated API key.

Implementing the home view

As expected, at this point it will be required of us to do a bit of coding. In keeping with practices we have adhered to in previous chapters, it will be wise of us to firstly make a barebones graphical mockup of the view we want to create before commencing with coding. This will save a significant amount of time in the long run by providing a clear direction regarding what we want to build.

We want the home page we are creating to do the following:

1. Show the latest place reviews posted on the platform
2. Provide direct access to a web page for review creation
3. Provide a means by which a user can sign out of their account
4. Enable a user to view the exact location of a reviewed place with the help of a map

Keeping all these requirements in mind, we can draw a rough sketch of our final template that looks something like this:

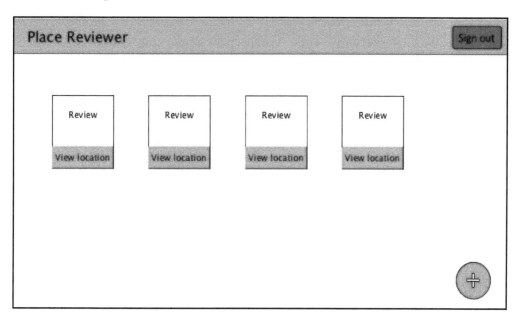

We are aiming for functionality over flash in our layout design. You can see that layout requirements one through four are immediately satisfied by this sketch. Clicking on **View location** will present the user of the application with a modal within which a map showing the exact place reviewed will be displayed.

Having clearly stated the template we will be creating, let us code. As always, first and foremost we need to include external stylesheets and scripts to our template. Open up `home.html` and add the following code to it:

```
<!DOCTYPE html>
<html lang="en" xmlns:th="http://www.thymeleaf.org">
  <head>
    <title>Home</title>
    <!-- Addition of external stylesheets -->
    <link rel="stylesheet" th:href="@{/css/app.css}"/>
    <link rel="stylesheet"
      href="https://cdnjs.cloudflare.com/ajax/libs/toastr.js
            /latest/toastr.min.css">
    <link rel="stylesheet"
      href="/webjars/bootstrap/4.0.0-beta.3/css/bootstrap.min.css"/>
    <link rel="stylesheet"
      href="https://maxcdn.bootstrapcdn.com/font-awesome
```

```
            /4.7.0/css/font-awesome.min.css"/>
    <link href="https://fastcdn.org/Buttons/2.0.0/css/buttons.css"
     rel="stylesheet">

    <!-- Inclusion of external Javascript -->
    <script src="/webjars/jquery/3.2.1/jquery.min.js"></script>
    <script src="https://cdnjs.cloudflare.com/ajax/libs
                 /toastr.js/latest/toastr.min.js">
    </script>
    <script src="https://cdnjs.cloudflare.com/ajax/libs
                 /popper.js/1.12.6/umd/popper.min.js">
    </script>
    <script src="/webjars/bootstrap/4.0.0-beta.3/
                 js/bootstrap.min.js"></script>
    <script src="https://fastcdn.org/Buttons/2.0.0/js/buttons.js"></script>
    <script type="text/javascript"
     src="https://maps.googleapis.com/maps/api/js?key={{API_KEY}}
            &libraries=places">
  </head>
</html>
```

Nice work! In addition to external stylesheets, we are going to make use of internal styles in this template. To define internal stylesheets in HTML files, simply add a `<style>` tag within the head of the HTML and input your desired CSS rules. Add the following style to `home.html`:

```
</script>

    <!-- Definition of internal styles -->
    <style>
      #map {
        height: 400px;
      }

      .container-review {
        background-color: white;
        border-radius: 2px;
        font-family:sans-serif;
        box-shadow: 0 0 1px black;
        border-color: black;
        padding: 0;
        min-width: 250px;
        height: 230px;
      }

      .review-author {
        font-size: 15px
```

```
    }

    .review-location {
      font-size: 12px
    }

    .review-title {
      font-size: 13px;
      text-decoration-style: dotted;
      height: calc(20 / 100 * 230px);
    }

    .review-content {
      font-size: 12px;
      height: calc(40 / 100 * 230px);
    }

    .review-header {
      height: calc(20 / 100 * 230px)
    }

    hr {
      margin: 0;
    }

    .review-footer {
      height: calc(20 / 100 * 230px);
    }
  </style>
```

Now, lets work on the body of the page. As you already know, all elements that make up the body of an HTML template must exist in a `<body>` tag. With that in mind, we can work on `home.html`. Start by adding the following HTML to the template file:

```
<!-- Invokes the showNoReviewNotification() function defined in -->
<!-- internal Javascript of this file upon document load. -->
<body
  th:onload="'javascript:showNoReviewNotification(
    ' + ${reviews.size() == 0} + '
  )'">
<div th:insert="fragments/navbar :: navbar"></div>
<div class="container">
  <div class="row mt-5">
    <!-- Creates view containers for each review retrieved -->
    <!-- Distinct <div> containers are created for the -->
    <!-- review author, location, title and body. -->
    <div th:each="review: ${reviews}"
```

```
          class="col-sm-2 container-review mt-4 mr-2">
           <div class="review-header pt-1">
              <div class="col-sm-12 review-author text-success">
                 <b th:text="${review.reviewer.username}"></b>
              </div>
              <div th:text="${review.placeName}"
               class="col-sm-12 review-location">
              </div>
           </div>
           <hr>
           <b>
              <div th:text="${review.title}"
               class="col-sm-12 review-title pt-1">
              </div>
           </b>
           <hr>
           <div th:text="${review.body}"
            class="col-sm-12 review-content pt-2">
           </div>
           <div class="review-footer">
              <!-- Creation of distinct DOM
              <button> elements for the display of reviewed locations. -->
                 <!-- Upon button click, the application renders a modal
                 showing the reviewed location on a map -->
                 <button class="col-sm-12 button button-small button-primary"
                 type="button" data-toggle="modal" data-target="#mapModal"
                 style="height: inherit; border-radius: 2px;"
                 th:onclick="'javascript:showLocation(
                    ' + ${review.latitude} + ','
                    + ${review.longitude} + ',\''
                    + ${review.placeId} + '\'
                 )'">
                 <i class="fa fa-map-o" aria-hidden="true"></i>
                 View location
              </button>
           </div>
        </div>
     </div>
  </div>
```

Great job! Do not worry too much about what the preceding code block does as of now. We shall explain everything in due time. Moving on continue the body of `home.html` by adding the following lines of code to it:

```
<!-- Modal creation -->
<div class="modal fade" id="mapModal">
   <div class="modal-dialog modal-lg" role="document">
```

```
<div class="modal-content">
  <div class="modal-header">
    <h5 class="modal-title">Reviewed location</h5>
    <button type="button" class="close"
     data-dismiss="modal" aria-label="Close">
      <span aria-hidden="true">&times;</span>
    </button>
  </div>
  <div class="modal-body">
    <div class="container-fluid">
      <div id="map"> </div>
    </div>
  </div>
  <div class="modal-footer">
    <button type="button" class="btn btn-primary"
     data-dismiss="modal">
      Done
    </button>
  </div>
</div>
</div>
</div>
```

The preceding code declares a modal that will hold a map displaying the exact location a review was created—upon the request of the user. We are not done with the home template yet. Continue work on the `<body>` further by adding the following lines of code to it:

```
<span style="bottom: 20px; right: 20px; position: fixed">
  <form method="get" th:action="@{/create-review}">
    <button class="button button-primary button-circle
     button-giant navbar-bottom" type="submit">
      <i class="fa fa-plus"></i>
    </button>
  </form>
</span>
```

Finally, add internal JavaScript for the HTML page:

```
<script>
  //Shows a toast notification to the user when no review is present
  function showNoReviewNotification(show) {
    if (show) {
      toastr.info('No reviews to see');
    }
  }
```

The following function initializes and displays a map showing the location where a review was written:

```
function showLocation(latitude, longitude, placeId) {
  var center = new google.maps.LatLng(latitude, longitude);

  var map = new google.maps.Map(document.getElementById('map'), {
    center: center,
    zoom: 15,
    scrollwheel: false
  });
  var service = new google.maps.places.PlacesService(map);

  loadPlaceMarker(service, map, placeId);
}
```

Load place marker creates a map marker on the reviewed location:

```
function loadPlaceMarker(service, map, placeId) {
  var request = {
    placeId: placeId
  };

  service.getDetails(request, function (place, status) {
    if (status === google.maps.places.PlacesServiceStatus.OK) {
      new google.maps.Marker({
        map: map,
        title: place.name,
      place: {
        placeId: place.place_id,
        location: place.geometry.location
      }

    })
    }
  });
}
    </script>
  </body>
```

Quite a lot has been done on `home.html`, so let's talk a bit about what exactly is going on in the view, starting with the `<head>` tag. We included stylesheets and scripts required by the home page from lines 4 through 16 of the `<head>` tag. The CSS included is as follows:

```
<link rel="stylesheet" th:href="@{/css/app.css}"/>
<link rel="stylesheet"
href="https://cdnjs.cloudflare.com/ajax/libs/toastr.js/latest/toastr.min.cs
s">
<link rel="stylesheet" href="/webjars/bootstrap/4.0.0-
beta.3/css/bootstrap.min.css"/>
<link rel="stylesheet"
href="https://maxcdn.bootstrapcdn.com/font-awesome/4.7.0/css/font-awesome.m
in.css"/>
<link href="https://fastcdn.org/Buttons/2.0.0/css/buttons.css"
rel="stylesheet">
```

These are external stylesheet inclusions for our application's CSS; Toastr, a library for the creation of JavaScript toast notifications; Bootstrap, a powerful library for designing websites and web applications; **Font Awesome**, an icon toolkit for websites and web applications; and buttons, a powerful and highly customizable web and CSS buttons library.

Right after the CSS inclusions, we have a number of external JavaScript inclusions as well:

```
<script src="/webjars/jquery/3.2.1/jquery.min.js"></script>
<script
src="https://cdnjs.cloudflare.com/ajax/libs/toastr.js/latest/toastr.min.js"
></script>
<script
src="https://cdnjs.cloudflare.com/ajax/libs/popper.js/1.12.6/umd/popper.min
.js"></script>
<script src="/webjars/bootstrap/4.0.0-beta.3/js/bootstrap.min.js"></script>
<script src="https://fastcdn.org/Buttons/2.0.0/js/buttons.js"></script>
<script type="text/javascript"
src="https://maps.googleapis.com/maps/api/js?key={{API_KEY}}&libraries=plac
es"></script>
```

The script inclusions in their respective order are for: JQuery, a JavaScript library designed specifically to simplify the client-side HTML scripting process; Toastr; Popper, a library used to manage poppers in web applications; Bootstrap; buttons; and the Google Places API web service. Once again, ensure you replace the `{{API_KEY}}` with your API key for the Google Places API web service—this is important.

Immediately after the JavaScript inclusions, we defined an internal stylesheet for the web page. Unfortunately, an explanation of stylesheets and their creation is beyond the scope of this book. However, it will be a good idea to brush up on CSS in your spare time. Further down `home.html`, we added a `<body>` tag as follows:

```
<body th:onload="'javascript:showNoReviewNotification(' + ${reviews.size()
== 0} + ')'">
```

`th:onload` in this is used to specify JavaScript that must be run after the page has been completely loaded. In short, it specifies code to be executed after an `onload` event occurs. In this case, the script to be run is a JavaScript function we defined further down the template, `showNoReviewNotification(boolean)`. The function shows a toast message indicating that no reviews are available to be viewed when the reviews list provided by the model is empty. `showNoReviewNotification(boolean)` is declared in our template as follows:

```
function showNoReviewNotification(show) {
  if (show) {
    toastr.info('No reviews to see');
  }
}
```

`showNoReviewNotification(boolean)` takes a single `Boolean` argument, `show`. When `show` is true, a toast notification with the message `No reviews to see` is rendered to the user. The display of toast notifications to a user is made possible by the Toastr library we are utilizing.

When there are reviews available to be shown to the user, then a container is created for each review item as follows:

```
<!-- Creates view containers for each review retrieved -->
<!-- Distinct <div> containers are created for the -->
<!-- review author, location, title and body. -->
<div th:each="review: ${reviews}" class="col-sm-2 container-review mt-4
mr-2">
  <div class="review-header pt-1">
    <div class="col-sm-12 review-author text-success">
      <b th:text="${review.reviewer.username}"></b>
```

```
    </div>
    <div th:text="${review.placeName}" class="col-sm-12 review-location">
    </div>
  </div>
  <hr>
  <b>
    <div th:text="${review.title}" class="col-sm-12 review-title pt-1">
    </div>
  </b>
  <hr>
  <div th:text="${review.body}" class="col-sm-12 review-content pt-2">
  </div>
  <div class="review-footer">
    <!-- Creation of distinct DOM <button> elements for the
      display of reviewed locations. -->
    <!-- Upon button click, the application renders a modal
      showing the reviewed location on a map -->
    <button class="col-sm-12 button button-small button-primary"
      type="button" data-toggle="modal" data-target="#mapModal"
      style="height: inherit; border-radius: 2px;"
      th:onclick="''javascript:showLocation('
        + ${review.latitude} + ','
        + ${review.longitude} + ',\''
        + ${review.placeId} + '\
      ')'">
      <i class="fa fa-map-o" aria-hidden="true"></i>
      View location
    </button>
  </div>
</div>
```

Each review container displays the username of the reviewer, the name of the place
reviewed, the review title, the review body and a button enabling the user to view the
reviewed location. Thymeleaf's `th:each` attribute was used to iterate over each review in
the `reviews` list, as shown here:

```
<div th:each="review: ${reviews}" class="col-sm-2 container-review mt-4
mr-2">
```

A good way to understand the iteration process is by reading `th:each="review:
${reviews}"` as `For each review in reviews`. The review currently being iterated
upon is held by the `review` variable. Hence, the data held by the review being iterated
upon can be accessed like any other object. This is the case here:

```
<div th:text="${review.placeName}" class="col-sm-12 review-location"></div>
```

`th:text` sets the text held by the `<div>` to the value assigned to `review.placeName`. It is also necessary to explain the process by which location maps are shown to the user. Take a close look at the following lines of code:

```
<button class="col-sm-12 button button-small button-primary" type="button"
  data-toggle="modal" data-target="#mapModal" style="height: inherit;
  border-radius: 2px;"
  th:onclick="'javascript:showLocation('
    + ${review.latitude} + ','
    + ${review.longitude} + ',\''
    + ${review.placeId} + '\
  ')'">
  <i class="fa fa-map-o" aria-hidden="true"></i>
  View location
</button>
```

This code block defines a button that does two things when a click event occurs on it. Firstly, it displays a modal identified by the ID `mapModal` to the user. Secondly, it initializes and renders a map displaying the exact location that was reviewed. The rendering of the map is made possible by the `showLocation()` JavaScript function we defined in our template file.

`showLocation()` takes three parameters as its arguments. The first is the longitudinal coordinate, the second a latitudinal coordinate, and the third the unique identifier of the location reviewed—a place ID. The place ID for the location is provided by the Google Places API. Firstly, `showLocation()` retrieves a central point for the locational coordinates provided. This is done by utilizing the Google Places API's `google.maps.LatLng` class. Simply defined, a `LatLng` is a point in geographical coordinates (longitude and latitude). Upon retrieving the central point, a new map is created with the use of the `Map` class (again, provided by the Google Places API) as shown in this code snippet:

```
var map = new google.maps.Map(document.getElementById('map'), {
  center: center,
  zoom: 15,
  scrollwheel: false
});
```

The created map is placed within a DOM container element with an ID `map`. After creating the necessary map, we create a location marker at the exact location with the help of the `loadPlaceMarker()` function. `loadPlaceMarker()` takes instances of `google.maps.places.PlacesService`, `Map`, and a place id as its three arguments. `PlacesService` is a class that possesses methods for the retrieval of place information and searching places.

The instance of `google.maps.places.PlacesService` is firstly used to retrieve the details of the place with the specified place ID (the reviewed location). If the details of the place are successfully retrieved, `status ===` `google.maps.places.PlacesServiceStatus.OK` evaluates to true and a marker for the location is placed on the map. The marker is created with the `google.maps.Marker` class. `Marker()` takes an optional options object as a its sole argument. When the options object is present, the place marker is created with the options specified. In this case, we specified a map in the options object. As such, the marker is added to the map upon its creation.

Lastly, we added a form to our template that sends a `GET` request to the `/reviews/new` path upon its submission, and added a button that submits the form upon clicking. This was done in the lines shown here:

```
<form method="get" th:action="@{/reviews/new}">
  <button class="button button-primary button-circle button-giant
    navbar-bottom" type="submit"><i class="fa fa-plus"></i></button>
</form>
```

That is all we need to do regarding the home page, so go ahead and check it out! Rebuild and run the application, register an account, and view the home page you just created!

As you can see, no reviews have been created on the platform for viewing. We must now work on a web page that allows the creation of reviews.

Implementing the review creation web page

Thus far, we have created views for user registration and login, as well as a homepage for logged in users to peruse reviews posted on the platform. We must now work on the view that facilitates the creation of these reviews. As always, first, before creating the view, let us work on an action that will be in charge of rendering our to-be-developed view to users. Open up the `ApplicationController` class and add the following method to it:

```kotlin
@GetMapping("/create-review")
fun createReview(model: Model, principal: Principal): String {
 model.addAttribute("principal", principal)
 return "create-review"
}
```

The `createReview()` action handles HTTP GET requests to the `/create-review` request path by returning a `create-review.html` template to the client for rendering. Go ahead and add a `create-review.html` file to the project `template` directory.

Similar to what we did before, let's begin by adding external styles and scripts to `create-review.html`:

```html
<!DOCTYPE html>
<html lang="en" xmlns:th="http://www.thymeleaf.org">
  <head>
    <title>New review</title>
    <!-- Addition of external stylesheets -->
    <link rel="stylesheet" th:href="@{/css/app.css}"/>
    <link rel="stylesheet" href="/webjars/bootstrap/4.0.0-beta.3
                                 /css/bootstrap.min.css"/>
    <link rel="stylesheet" href="https://maxcdn.bootstrapcdn.com
                                 /font-awesome/4.7.0/css/font-awesome.min.css"/>
    <link href="https://fastcdn.org/Buttons/2.0.0/css/buttons.css"
     rel="stylesheet">

    <!-- Inclusion of external Javascript -->
    <script src="/webjars/jquery/3.2.1/jquery.min.js"></script>
    <script src="https://cdnjs.cloudflare.com/ajax/libs/popper.js
                /1.12.6/umd/popper.min.js"></script>
    <script src="/webjars/bootstrap/4.0.0-beta.3/
                js/bootstrap.min.js"></script>
    <script src="https://fastcdn.org/Buttons/2.0.0/js/buttons.js"></script>
    <script type="text/javascript" src="https://maps.googleapis.com/
                maps/api/js?key={{API_KEY}}&libraries=places">
    </script>
```

Now, add our required internal stylesheet for the webpage:

```
<!-- Definition of internal styles -->
<style>
  #map {
    height: 400px;
  }

  #container-place-data {
    height: 0;
    visibility: hidden;
  }

  #container-place-info {
    font-size: 14px;
  }

  #container-selection-status {
    visibility: hidden;
  }
</style>
</head>
```

The next thing on our agenda is the creation of the necessary form for the input of review data. Continue the `create-review.html` template with the following code:

```
<body>
  <div th:insert="fragments/navbar :: navbar"> </div>
  <div class="container-fluid">
    <div class="row">
      <div class="col-sm-12 col-xs-12">
        <!-- Review form creation -->
        <form class="form-group col-sm-12 form-vertical form-app"
         id="form-login" method="post" th:action="@{/reviews}">
          <div class="col-sm-12 mt-2 lead">Write your review</div>
          <div th:if="${error != null}" class="text-danger"
           th:text="${error}"> </div>
          <hr>
          <input class="form-control" type="text" name="title"
           placeholder="Title" th:value="${title}" required/>
          <textarea class="form-control mt-4" rows="13" name="body"
           placeholder="Review" th:value="${body}" required></textarea>
          <div class="form-group" id="container-place-data">
            <!-- Input fields for location specific form data -->
            <!-- Form input data for the fields below are
             provided by the Google Places API -->
            <input class="form-control" id="place_address"
```

```
       th:value="${placeAddress}" type="text" name="placeAddress"
       required/>
      <input class="form-control" id="place_name" type="text"
      name="placeName" th:value="${placeName}" required/>
      <input class="form-control" id="place_id" type="text"
      name="placeId" th:value="${placeId}" required/>
      <input id="location-lat" type="number" name="latitude"
      step="any" th:value="${latitude}" required/>
      <input id="location-lng" type="number" name="longitude"
      step="any" th:value="${longitude}" required/>
    </div>
    <div class="form-group mb-3">
      <button class="button button-pill" type="button"
      data-toggle="modal" data-target="#mapModal">
        <i class="fa fa-map-marker" aria-hidden="true"></i>
        Select Location
      </button>
      <button class="button button-pill button-primary">
      Submit Review</button>
      </div>
      <div class="text-success ml-2" id="container-selection-status">
      Location selected</div>
    </form>
  </div>
</div>
```

Now, let us add a modal that will enable the user to select the review location from a map. Do not worry too much about the details of the selection process as of now. We shall talk more on it shortly:

```
<!-- Map Modal -->
<div class="modal fade" id="mapModal">
  <div class="modal-dialog modal-lg" role="document">
    <div class="modal-content">
      <div class="modal-header">
        <h5 class="modal-title">Select place to review</h5>
        <button type="button" class="close" data-dismiss="modal"
        aria-label="Close">
          <span aria-hidden="true">&times;</span>
        </button>
      </div>
      <div class="modal-body">
        <div class="container-fluid">
          <div id="map"> </div>
            <div class="row mt-2" id="container-place-info">
              <div class="col-sm-12" id="container-place-name">
                <b>Place Name:</b>
```

```
          </div>
          <div class="col-sm-12" id="container-place-address">
            <b>Place Address:</b>
          </div>
        </div>
      </div>
    </div>
    <div class="modal-footer">
      <button type="button" class="btn btn-primary"
        data-dismiss="modal">Done</button>
    </div>
  </div>
 </div>
 </div>
</div>
```

And lastly, we finish up the template by including its internal JavaScript, as shown in the code snippets that follow:

```
<script>
  // form field reference creation
  var formattedAddressField = document
                       .getElementById('place_address');
  var placeNameField = document.getElementById('place_name');
  var placeIdField = document.getElementById('place_id');
  var latitudeField = document.getElementById('location-lat');
  var longitudeField = document.getElementById('location-lng');

  // container reference creation
  var containerPlaceName = document.getElementById
                            ('container-place-name');
  var containerPlaceAddress = document.getElementById
                            ('container-place-address');
  var containerSelectionStatus = document.getElementById
                            ('container-selection-status');
```

In the preceding code snippet, we created references to important DOM elements that exist on the page. These references include references to place specific input fields (the fields for the address, name, ID, and latitudinal and longitudinal coordinates of a place). In addition, we added references for the containers displaying the details of a selected place, such as the place name and place address. At this juncture, we will declare a few functions. These functions are `initialize()`, `getPlaceDetailsById()`, `updateViewData()`, `setFormValues()`, `showSelectionsStatusContainer()`, and `setContainerText()`.

Start by adding the `initialize()` and `getPlaceDetailsById()` functions shown here to the template:

```
//invoked to initialize Google map
function initialize() {

    navigator.geolocation.getCurrentPosition(function(location) {
        var latitude = location.coords.latitude;
        var longitude = location.coords.longitude;

        var center = new google.maps.LatLng(latitude, longitude);

        var map = new google.maps.Map(document.getElementById('map'), {
            center: center,
            zoom: 15,
            scrollwheel: false
        });

        var service = new google.maps.places.PlacesService(map);

        map.addListener('click', function(data) {
            getPlaceDetailsById(service, data.placeId);
        });
    });

}
```

We have added the function below to enable us to get the details of a particular place from the Google Places API Invoked to retrieve the details of a place:

```
function getPlaceDetailsById(service, placeId) {
    var request = {
        placeId: placeId
    };

    service.getDetails(request, function (place, status) {
        if (status === google.maps.places.PlacesServiceStatus.OK) {
            updateViewData(place)
        }
    });
}
```

Now, add `updateView()` and `setFormValues()` as shown here:

```
//Invoked to update view information
function updateViewData(place) {
  setFormValues(
    place.formatted_address,
    place.name,
    place.place_id,
    place.geometry.location.lat(),
    place.geometry.location.lng()
  );

  setContainerText('<b>Place Name: </b>' + place.name,
    '<b>Place Address: </b>' + place.formatted_address);

  showSelectionStatusContainer();
}
```

The function below is called to update view form data:

```
function setFormValues(formattedAddress, placeName, placeId,
                       latitude, longitude) {
  formattedAddressField.value = formattedAddress;
  placeNameField.value = placeName;
  placeIdField.value = placeId;
  latitudeField.value = latitude;
  longitudeField.value = longitude;
}
```

Finally, finish up the template by adding the code shown here:

```
function showSelectionStatusContainer() {
  containerSelectionStatus.style.visibility = 'visible'
}

function setContainerText(placeNameText, placeAddressText) {
  containerPlaceName.innerHTML = placeNameText;
  containerPlaceAddress.innerHTML = placeAddressText;
}

// Initializes map upon window load completion
google.maps.event.addDomListener(window, 'load', initialize);
    </script>
  </body>
</html>
```

Similar to the previous template, we begin `create-review.html` with the addition of both external and internal CSS and JavaScript required by the template in the HTML `<head>` tag. Further into the template, we create a form that takes the following form data as its input:

- `title`: A user defined title for the review being created.
- `body`: The body of the review. This is the main review text.
- `placeAddress`: The address of the place being reviewed.
- `placeName`: The name of the place being reviewed.
- `placeId`: The unique ID of the location being reviewed.
- `latitude`: The latitudinal coordinate of the reviewed location.
- `longitude`: The longitudinal coordinate of the reviewed location.

There will be no need for the user to provide form input for `placeAddress`, `placeName`, `placeId`, `latitude`, and `longitude`. As such, we have hidden the parent `<div>` of the aforementioned input elements. We shall utilize the Google Places API to retrieve place-specific information. Make sure to note that in the template we are using a modal to display a map for location selection. The modal is toggled by a `button` we added in our template as follows:

```
<button class="button button-pill" type="button" data-toggle="modal" data-target="#mapModal">
  <i class="fa fa-map-marker" aria-hidden="true"></i> Select Location
</button>
```

Clicking on the button will display the map modal to the user. Upon rendering the map, a user can click on their desired review location from the map. Performing such a click action will trigger the map's click event, which in turn will be handled by the listener we defined in the template as follows:

```
map.addListener('click', function(data) {
  getPlaceDetailsById(service, data.placeId);
});
```

`getPlacesDetailsById()` takes two arguments: an instance of
`google.maps.places.PlacesService` and the ID of the place whose information is to be
retrieved. The `PlacesService` instance is then used to retrieve the information of the
place. After this information retrieval, the view is duly updated with the information
retrieved: place-specific form data is set, the place name and address container within the
map modal is updated, and a message indicating that a location has successfully been
selected is shown to the user. Upon selection of a location and the input of all required form
data, the user can then submit their review.

We are almost ready to try out the review creation page. Before we do, we must create a
review validator, as well as a controller action that handles POST requests sent to the
`/reviews` path. Let's start with `ReviewValidator`. Add the `ReviewValidator` class
shown here to `com.example.placereviewer.component`:

```
package com.example.placereviewer.component

import com.example.placereviewer.data.model.Review
import org.springframework.stereotype.Component
import org.springframework.validation.Errors
import org.springframework.validation.ValidationUtils
import org.springframework.validation.Validator

@Component
class ReviewValidator: Validator {

  override fun supports(aClass: Class<*>?): Boolean {
    return Review::class == aClass
  }

  override fun validate(obj: Any?, errors: Errors) {
    val review = obj as Review

    ValidationUtils.rejectIfEmptyOrWhitespace(errors, "title",
                        "Empty.reviewForm.title", "Title cannot be empty")
    ValidationUtils.rejectIfEmptyOrWhitespace(errors, "body",
                        "Empty.reviewForm.body", "Body cannot be empty")
    ValidationUtils.rejectIfEmptyOrWhitespace(errors, "placeName",
                        "Empty.reviewForm.placeName")
    ValidationUtils.rejectIfEmptyOrWhitespace(errors, "placeAddress",
                        "Empty.reviewForm.placeAddress")
    ValidationUtils.rejectIfEmptyOrWhitespace(errors, "placeId",
                        "Empty.reviewForm.placeId")
    ValidationUtils.rejectIfEmptyOrWhitespace(errors, "latitude",
                        "Empty.reviewForm.latitude")
    ValidationUtils.rejectIfEmptyOrWhitespace(errors, "longitude",
```

```
                          "Empty.reviewForm.longitude")

    if (review.title.length < 5) {
      errors.rejectValue("title", "Length.reviewForm.title",
                         "Title must be at least 5 characters long")
    }

    if (review.body.length < 5) {
      errors.rejectValue("body", "Length.reviewForm.body",
                         "Body must be at least 5 characters long")
    }
  }
}
```

As we have previously explained the workings of custom validators, there is little need to explain how this validator works. Without taking time, let us implement a controller class for HTTP requests pertaining to reviews. Create a `ReviewController` class in `com.example.placereviewer.controller` and add the following code to it:

```
package com.example.placereviewer.controller

import com.example.placereviewer.component.ReviewValidator
import com.example.placereviewer.data.model.Review
import com.example.placereviewer.service.ReviewService
import org.springframework.stereotype.Controller
import org.springframework.ui.Model
import org.springframework.validation.BindingResult
import org.springframework.web.bind.annotation.ModelAttribute
import org.springframework.web.bind.annotation.PostMapping
import org.springframework.web.bind.annotation.RequestMapping
import javax.servlet.http.HttpServletRequest

@Controller
@RequestMapping("/reviews")
class ReviewController(val reviewValidator: ReviewValidator,
                       val reviewService: ReviewService) {

  @PostMapping
  fun create(@ModelAttribute reviewForm: Review, bindingResult:
BindingResult,
             model: Model, request: HttpServletRequest): String {
    reviewValidator.validate(reviewForm, bindingResult)

    if (!bindingResult.hasErrors()) {
      val res = reviewService.createReview(request.userPrincipal.name,
                                           reviewForm)
```

```
    if (res) {
      return "redirect:/home"
    }
  }

  with (model) {
    addAttribute("error", bindingResult.allErrors.first().defaultMessage)
    addAttribute("title", reviewForm.title)
    addAttribute("body", reviewForm.body)
    addAttribute("placeName", reviewForm.placeName)
    addAttribute("placeAddress", reviewForm.placeAddress)
    addAttribute("placeId", reviewForm.placeId)
    addAttribute("longitude", reviewForm.longitude)
    addAttribute("latitude", reviewForm.latitude)
  }

  return "create-review"
  }
}
```

Having added the `ReviewValidator` and `ReviewController` classes, build and run the project, log in as a user, and navigate to `http://localhost:5000/create-review` from your favorite browser.

Upon page load, you will be presented with a form you can use to add a new review:

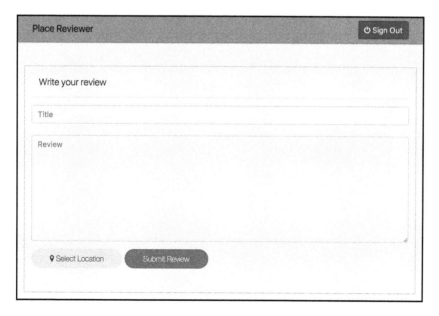

Users are required to select a review location before a review can be submitted. To select a review location, click the **Select Location** button:

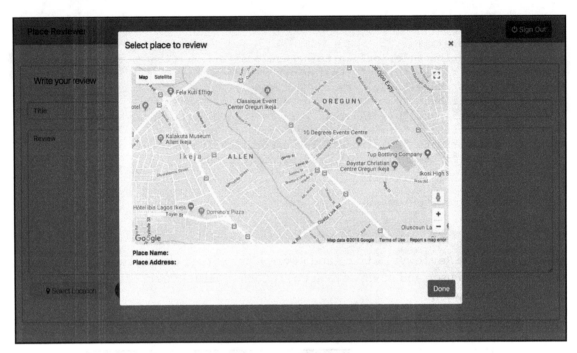

Clicking the **Select Location** button will present the user with a modal containing a map from which they can select a location of choice to review. Clicking on a location from the map will bring up an information window on the map containing data pertaining to the clicked location. In addition, the modal container for holding the selected place name and address will be updated:

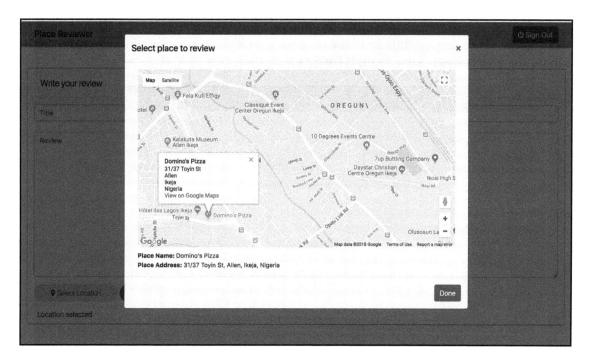

After a user selects their review location of choice, they can close the modal by clicking **Done** and proceed to filling in the title and body of their review:

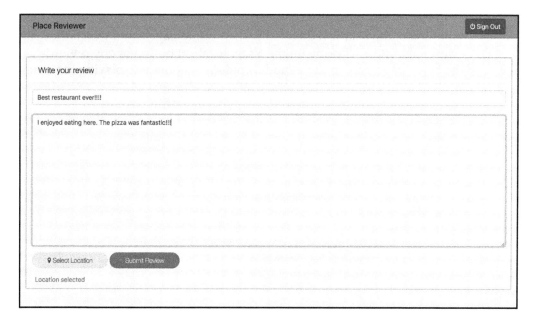

Notice that the review form now indicates that a review location has been selected successfully. Once the user fills in all the necessary review information, they can proceed to submit the review by clicking **Submit Review**:

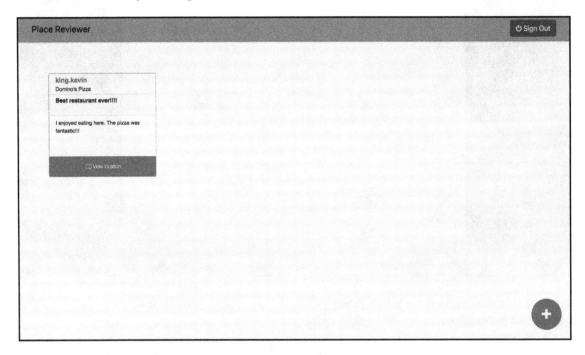

After review submission, the user is redirected to their home page, from which they can now see the review submitted. Clicking on the **View location** button of any review displayed on the home page will render a modal containing a map displaying the exact location reviewed:

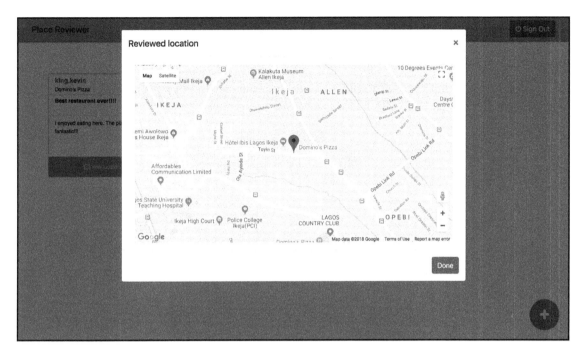

The map displayed to the user possesses a marker indicating the exact location reviewed by the reviewer.

At this stage, we have concluded all the core functionality of the *Place Reviewer* application. Before we wrap up this chapter, let us explore how to test Spring applications.

Spring application testing

Earlier on in this book, we discussed application testing and why it is necessary in creating reliable software. We must now explore the process of testing Spring applications. A Spring application can be tested in four easy steps:

1. Add necessary testing dependencies to the project
2. Create a configuration class
3. Configure test class to use custom configuration
4. Write required tests

We shall look at each of these steps one by one.

Adding necessary testing dependencies to the project

This involves the inclusion of suitable testing dependencies in your project. Open the `Place Reviewer` project's `pom.xml` file and add the following dependencies:

```xml
<dependency>
  <groupId>junit</groupId>
  <artifactId>junit</artifactId>
  <version>4.12</version>
  <scope>test</scope>
  <exclusions>
    <exclusion>
      <groupId>org.hamcrest</groupId>
      <artifactId>hamcrest-core</artifactId>
    </exclusion>
  </exclusions>
</dependency>
<dependency>
  <groupId>org.hamcrest</groupId>
  <artifactId>hamcrest-library</artifactId>
  <version>1.3</version>
  <scope>test</scope>
</dependency>
```

In the coming sections, we are going to learn how to write tests with jUnit and Hamcrest. JUnit is a testing framework for the Java programming language and Hamcrest is a library that provides matchers, which can be combined to create meaningful expressions of intent.

Creating a configuration class

The creation of a test configuration class aids the proper running of written tests. Within the `src/test/kotlin` directory of the `Place Reviewer` project, add a `config` package to `com.example.placereviewer`. Add the `TestConfig` class shown here to the package created:

```kotlin
package com.example.placereviewer.config

import org.springframework.context.annotation.ComponentScan
import org.springframework.context.annotation.Configuration

@Configuration
@ComponentScan(basePackages = ["com.example.placereviewer"])
```

```
class TestConfig
```

Configuring a test class to use custom configuration

To do this, open the Spring application's test class and use an `@ContextConfiguration` annotation to specify the configuration classes to be used by the test class. Open `PlaceReviewerApplicationTests.kt` (located in the `com.example.placereviewer` package of your project's `src/test/kotlin` directory). Now, set its configuration class as follows:

```
package com.example.placereviewer

import com.example.placereviewer.config.TestConfig
import org.junit.runner.RunWith
import org.springframework.boot.test.context.SpringBootTest
import org.springframework.test.context.ContextConfiguration
import org.springframework.test.context.junit4.SpringRunner

@RunWith(SpringRunner::class)
@SpringBootTest
@ContextConfiguration(classes = [TestConfig::class])
class PlaceReviewerApplicationTests
```

Great work! You are now ready to write some application tests.

Writing your first test

Writing code for application tests is like writing code for any other part of a Spring application. You can make use of components and services as you would in any other part of your application. Let us demonstrate this, shall we?

Add the following `TestUserService` interface to `com.example.placereviewer.service`:

```
package com.example.placereviewer.service

import com.example.placereviewer.data.model.User

interface TestUserService {
  fun getUser(): User
}
```

Now, add the following `TestUserServiceImpl` class to the package:

```
package com.example.placereviewer.service

import com.example.placereviewer.data.model.User
import org.springframework.stereotype.Service

@Service
internal class TestUserServiceImpl : TestUserService {

  //Test stub mimicking user retrieval
  override fun getUser(): User {
    return User(
      "user@gmaiil.com",
      "test.user",
      "password"
    )
  }
}
```

Return to the `PlaceReviewerApplicationTests.kt` file and modify it to reflect these changes:

```
package com.example.placereviewer

import com.example.placereviewer.config.TestConfig
import com.example.placereviewer.data.model.User
import com.example.placereviewer.service.TestUserService
import org.hamcrest.Matchers.instanceOf
import org.hamcrest.MatcherAssert.assertThat
import org.junit.Test
import org.junit.runner.RunWith
import org.springframework.beans.factory.annotation.Autowired
import org.springframework.boot.test.context.SpringBootTest
import org.springframework.test.context.ContextConfiguration
import org.springframework.test.context.junit4.SpringRunner

@RunWith(SpringRunner::class)
@SpringBootTest
@ContextConfiguration(classes = [TestConfig::class])
class PlaceReviewerApplicationTests {

  @Autowired
  lateinit var userService: TestUserService

  @Test
  fun testUserRetrieval() {
```

```
    val user = userService.getUser()

    assertThat(user, instanceOf(User::class.java))
  }
}
```

The `testUserRetrieval()` method is a test that, when run, makes use of the stub method we defined in `TestUserServiceImpl` to retrieve a user and asserts that the object returned by the function is an instance of the `User` class.

To run a written test, click the **Run Test** button to the right of the created test in the IDE window:

testUserRetrieval will be run and the result of the test run will be displayed at the bottom of the IDE window:

```
                            1 test passed - 467ms
2018-03-04 22:09:52.292  INFO 39684 --- [          main] s.w.s.m.m.a.RequestMappingHandlerMapping  : Mapped "{[/error],produces=[tex
2018-03-04 22:09:52.322  INFO 39684 --- [          main] o.s.w.s.handler.SimpleUrlHandlerMapping   : Mapped URL path [/login] onto h
2018-03-04 22:09:52.352  INFO 39684 --- [          main] o.s.w.s.handler.SimpleUrlHandlerMapping   : Mapped URL path [/webjars/**] o
2018-03-04 22:09:52.353  INFO 39684 --- [          main] o.s.w.s.handler.SimpleUrlHandlerMapping   : Mapped URL path [/css/**] onto
2018-03-04 22:09:52.353  INFO 39684 --- [          main] o.s.w.s.handler.SimpleUrlHandlerMapping   : Mapped URL path [/**] onto hand
2018-03-04 22:09:52.403  WARN 39684 --- [          main] o.s.h.c.j.Jackson2ObjectMapperBuilder     : For Jackson Kotlin classes supp
2018-03-04 22:09:52.485  INFO 39684 --- [          main] o.s.w.s.handler.SimpleUrlHandlerMapping   : Mapped URL path [/**/favicon.ic
2018-03-04 22:09:52.675  WARN 39684 --- [          main] o.s.h.c.j.Jackson2ObjectMapperBuilder     : For Jackson Kotlin classes supp
2018-03-04 22:09:53.001  INFO 39684 --- [          main] c.e.p.PlaceReviewerApplicationTests       : Started PlaceReviewerApplicatio
2018-03-04 22:09:53.495  INFO 39684 --- [       Thread-3] o.s.w.c.s.GenericWebApplicationContext    : Closing org.springframework.web
2018-03-04 22:09:53.500  INFO 39684 --- [       Thread-3] j.LocalContainerEntityManagerFactoryBean  : Closing JPA EntityManagerFactor
2018-03-04 22:09:53.504  INFO 39684 --- [       Thread-3] com.zaxxer.hikari.HikariDataSource        : testdb - Shutdown initiated...
2018-03-04 22:09:53.516  INFO 39684 --- [       Thread-3] com.zaxxer.hikari.HikariDataSource        : testdb - Shutdown completed.

Process finished with exit code 0
```

In this case, the test we wrote passed. That's a great thing. However, as you develop larger and more complex applications and write tests for application modules, you will discover that more often than not, written tests fail. When this happens, do not fret; simply stay calm and debug your application. As time passes, you will learn to create more reliable software.

Summary

In this chapter, we wrapped up our journey through Kotlin by finishing up the *Place Reviewer* application. In the process, we explored—in depth—the creation of view layers for Spring MVC-based applications. Furthermore, we learned how to integrate an application with Google Places API Web Services with the goal of making an application location-aware.

In addition, we learned about form input validation with the help of Validator classes and BindingResult. Finally, we covered how to configure for testing and write tests for Spring-based applications.

What next?

If you are reading this section, you have most likely successfully gone through this book. First of all, allow me to congratulate you on making it this far! Your dedication and drive towards learning Kotlin is nothing short of praiseworthy. At this stage, the question *what next?* is likely running through your mind. This section was made to specifically to answer that question.

First and foremost, it is necessary that you strive to attain mastery of Kotlin. This will take dedication and practice—lots and lots of practice—but ultimately it is doable. The following will help you get to the level of mastery that you desire:

- **Practice programming with Kotlin daily**: This is highly important as it will ensure that what you have learned about Kotlin so far sticks. In addition, doing this will help you discover new constructs, patterns, data structures, and paradigms that will take your programming skills to the next level.
- **Read! Read! Read!**: This cannot be emphasized enough. In order to attain mastery of any skill, it is important that you acquire as much knowledge as possible pertaining to it. As such, you must make reading about Kotlin and Kotlin-related topics a habit. An amazing place to start is the official Kotlin reference, which you can access at this link: `https://kotlinlang.org/docs/reference`. Furthermore, do not shy away from picking up other quality books on Kotlin.
- **Ask and answer questions relating to Kotlin**: At many points in your journey to mastery of Kotlin, questions about the language—both big and small—will pop into your mind. Do not let these questions go unanswered. If you do, you will be missing out on a beautiful learning experience. Platforms such as Stack Overflow and Quora were made specifically for knowledge sharing. Make sure you form a habit of using these platforms to get answers to your most pressing questions about the language. Also, feel free to answer questions when you are comfortable with doing so.
- **Build things!**: This goes without saying. Few things give more experience with a tool than building with the tool yourself. Make sure you do not shy away from taking up offers to build Kotlin applications. Building is necessary for rapid progress with a tool.

If you maintain a passion for Kotlin and follow these suggestions, you will attain mastery of Kotlin in no time. Best of luck with your journey ahead!

Other Books You May Enjoy

If you enjoyed this book, you may be interested in these other books by Packt:

Functional Kotlin
Mario Arias, Rivu Chakraborty

ISBN: 978-1-78847-648-5

- Learn the Concepts of Functional Programming with Kotlin
- Discover the Coroutines in Kotlin
- Uncover Using funkTionale plugin
- Learn Monads, Functiors and Applicatives
- Combine Functional Programming with OOP and Reactive Programming
- Uncover Using Monads with funkTionale
- Discover Stream Processing

Kotlin Blueprints

Ashish Belagali, Hardik Trivedi, Akshay Chordiya

ISBN: 978-1-78839-080-4

- See how Kotlin's power and versatility make it a great choice to create applications across various platforms, and how it delivers business and technology benefits
- Write a robust web applications using Kotlin with Spring Boot
- Write Android applications with ease using Kotlin
- Write rich desktop applications in Kotlin
- Learn how Kotlin can generate Javascript and how this can be used on client side and server side development
- Understand how native applications can be written with Kotlin/Native
- Learn the practical aspects of programming in each of the applications

Leave a review - let other readers know what you think

Please share your thoughts on this book with others by leaving a review on the site that you bought it from. If you purchased the book from Amazon, please leave us an honest review on this book's Amazon page. This is vital so that other potential readers can see and use your unbiased opinion to make purchasing decisions, we can understand what our customers think about our products, and our authors can see your feedback on the title that they have worked with Packt to create. It will only take a few minutes of your time, but is valuable to other potential customers, our authors, and Packt. Thank you!

Index

www.ingramcontent.com/pod-product-compliance
Lightning Source LLC
Chambersburg PA
CBHW060641060326
40690CB00020B/4475